Crystalline Materials
for
Actinide Immobilisation

Series on Materials for Engineering

Editor-in-Chief: William (Bill) E. Lee (*Imperial College London, UK*)

Vol. 1:   Crystalline Materials for Actinide Immobilisation
          by B. E. Burakov, M. I. Ojovan and W. E. Lee

MATERIALS FOR ENGINEERING Vol. 1

Crystalline Materials
for
Actinide Immobilisation

Boris E. Burakov (V G Khlopin Radium Institute, Russia)

Michael I. Ojovan (University of Sheffield, UK)

William (Bill) E. Lee (Imperial College London, UK)

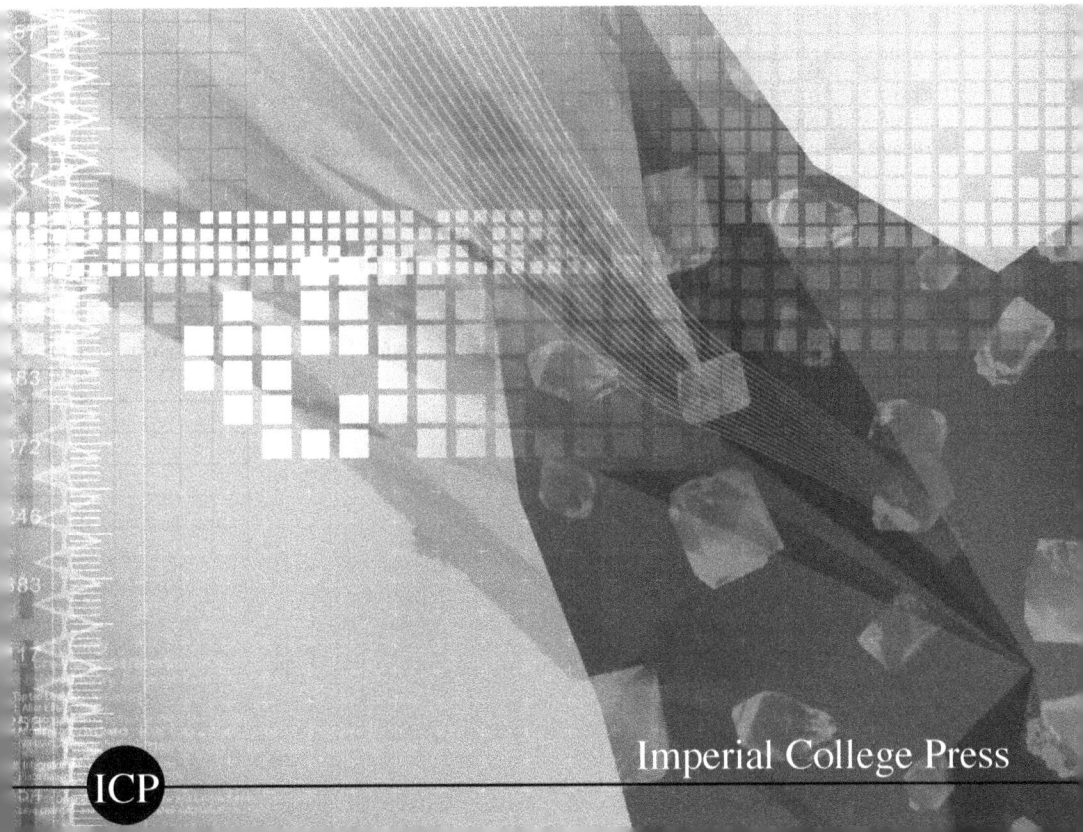

Imperial College Press

ICP

Published by

Imperial College Press
57 Shelton Street
Covent Garden
London WC2H 9HE

Distributed by

World Scientific Publishing Co. Pte. Ltd.
5 Toh Tuck Link, Singapore 596224
USA office: 27 Warren Street, Suite 401-402, Hackensack, NJ 07601
UK office: 57 Shelton Street, Covent Garden, London WC2H 9HE

British Library Cataloguing-in-Publication Data
A catalogue record for this book is available from the British Library.

CRYSTALLINE MATERIALS FOR ACTINIDE IMMOBILISATION
Materials for Engineering — Vol. 1

ISBN-13 978-1-84816-418-5
ISBN-10 1-84816-418-1

Typeset by Stallion Press
Email: enquiries@stallionpress.com

Printed in Singapore.

In memory of Georgiy Alexeevich Ilyinskiy, Professor of Mineralogy,
Leningrad (St. Petersburg) State University (1928–1997)

Preface

This book summarises approaches and current practices in actinide immobilisation using chemically-durable crystalline materials such as polycrystalline ceramics and large single crystals. Durable actinide-containing materials have many potential applications, such as in nuclear fuels for burning excess Pu and in chemically-inert sources of alpha irradiation for power supply of unmanned space vehicles or to produce electricity for microelectronic devices. However, currently these elements are considered mostly as waste constituents and actinide-bearing wastes have accumulated in various countries as a result of nuclear weapons' production. Excess weapon and civil Pu from commercial spent fuel awaits environmentally-safe immobilisation. Actinides are chemical elements with unique features that could be beneficially used in different areas of human endeavour including medicine. At present there is no appropriate balance between safe actinide use and disposal and there are difficult ethical questions about their use. Both use and disposal of actinides require their immobilisation in a durable host material. The choice of an optimal actinide immobilisation route is a significant technical challenge. While there is a wealth of information about actinide properties in many publications, there is a dearth of reviews or books summarising the current approach to actinide immobilisation. This book hopes to fill the gap based on the authors' experience and studies in nuclear material management and actinide immobilisation.

The first chapter introduces the actinide series with a brief description of basic physical and chemical properties emphasising the hazards associated with actinides. Natural actinides and actinide-bearing minerals found in nature are described. Artificial actinides from the nuclear fuel cycle are outlined, along with the most durable crystalline host-phases for them. Chapter Two gives a description of areas of actinide use such as in sealed sources of radiation, transmutation targets and advanced nuclear fuels. Nuclear waste actinide immobilisation is examined in the third chapter with emphasis on the most suitable host-matrices such as Synroc and other ceramics. The core of the book forms Chapters Four and Five, which give detailed descriptions of synthesis methods used to produce chemically-durable crystalline materials containing actinides and practical methods for their analysis. Radiation damage is an important issue for actinides and is examined in Chapter Six, where both ion bombardment and doping with shorter-lived radionuclides methods are analysed. The book concludes with a brief description of the future potential of actinide-containing materials.

This book is intended for nuclear waste management experts, radiochemists, geochemists, geologists, nuclear physicists, materials scientists and engineers, solid state physicists and cancer treatment experts. It will also be useful for a broader range of specialists interested in environmentally-safe use of radionuclides.

B. E. Burakov, M. I. Ojovan and W. E. Lee

Abbreviations

ANSTO	—	Australian Nuclear Science and Technology Organization
CCM	—	Cold crucible melting
CL	—	Cathodoluminescence
EPMA	—	Electron probe microanalysis
FP	—	Fission product
GGG	—	Gadolinium-gallium garnet
HLW	—	High-level radioactive wastes
HRTEM	—	High resolution transmission electron microscopy
IAEA	—	International Atomic Energy Agency
ICPP	—	Idaho Chemical Processing Plant at the Idaho National Engineering Laboratory (INEL), USA
IGEM	—	Institute of Geology of Ore Deposits, Moscow, Russia
IMF	—	Inert matrix fuel
ISL	—	Immobilisation Science Laboratory, Sheffield, UK
ITU	—	Institute for Transuranium Elements, Karlsruhe, Germany
JAERI	—	Japan Atomic Energy Research Institute
KRI	—	V.G. Khlopin Radium Institute, St. Petersburg, Russia
LANL	—	Los Alamos National Laboratory, NM, USA
LLNL	—	Lawrence Livermore National Laboratory, CA, USA
LWR	—	Light-water reactor

MA	—	Minor actinide
MOX	—	Mixed oxide nuclear fuel
NMR	—	Nuclear magnetic resonance
NZP	—	Sodium zirconium phosphate
ORNL	—	Oak Ridge National Laboratory, USA
P&C	—	Partitioning and conditioning
P&T	—	Partitioning and transmutation
PNNL	—	Pacific Northwest National Laboratory, Richland, Washington, USA
PSU	—	Pennsylvania State University, USA
PWR	—	Pressurized water reactor
RBMK	—	Channel-type uranium-graphite reactor
REE	—	Rare earth elements
RIAR	—	Research Institute of Atomic Reactors, Dimitrovgrad, Russia
ROSATOM	—	Russian State Corporation for Nuclear Energy
RTG	—	Radioisotope thermoelectric generator
SEM	—	Scanning electron microscopy
SIA "Radon"	—	Scientific and Industrial Association "Radon", Moscow, Russia
SNF	—	Spent nuclear fuel
SRS	—	Sealed radioactive source
TEM	—	Transmission electron microscopy
TEOS	—	Tetraethoxysilane
TNT	—	Trinitrotolouene
TPD	—	Thorium phosphate diphosphate
UKAEA	—	United Kingdom Atomic Energy Authority
XRD	—	X-ray diffraction
VNIINM	—	Institute of Inorganic Materials, Moscow, Russia
YAG	—	Yttrium-aluminium garnet

Acknowledgements

This book is dedicated to the memory of Professor Georgiy Alexee-vich Ilyinskiy, who was teaching mineralogy and was the scientific supervisor of one of authors (B. E. Burakov) when he was a student in Leningrad (now St. Petersburg) State University. The authors acknowledge with thanks the help of KRI colleagues Vladimir Zirlin, Larisa Nikolaeva, Elena Strykanova, Marina Petrova, Vladimir Garbuzov, Alexander Kitsay and Evgeniy Shashukov, as well as that of colleagues from Ioffe Physico-Technical Institute, including Maria Zamoryanskaya, Yana Domracheva, Katerina Kolesnikova and Maria Yagovkina. We appreciate cooperation with Sergey Yudintsev of the Institute of Geology of Ore Deposits (IGEM) and are thankful to Sergey Krivovichev and Roman Bogdanov from St. Petersburg State University, to Nicolas Dacheux of Institut de Chimie Séparative de Marcoule, Université de Montpellier and to Victoria Gribova of Nevsky Institute of Language and Culture for useful information. The authors are most grateful to colleagues at the ISL of the University of Sheffield including Martin Stennett, Neil Hyatt and John Roberts, to Ian Donald of AWE, Francis Livens of the University of Manchester, Robin Grimes and Alexander Chroneos of Imperial College London, Lou Vance of ANSTO, the late Igor A. Sobolev, Sergey A. Dmitriev, Sergey Stefanovky, Vsevolod Klimov, Olga Batyukhnova and Tatyana Scherbatova of Moscow SIA "Radon".

Contents

Preface   vii

Abbreviations   ix

Acknowledgements   xi

1. Introduction to the Actinides   1

    1.1   Actinide Series   1
        1.1.1   History   1
        1.1.2   Basic physical and chemical properties . . .   8
        1.1.3   History of using actinide-containing
                 materials   8
        1.1.4   High toxicity and long-lived radioactivity . .   17
        1.1.5   Need for actinide immobilisation   19
    1.2   Natural Actinides and Minerals   21
        1.2.1   Uraninite, pitchblende and thorianite   21
        1.2.2   Coffinite and thorite   23
        1.2.3   Brannerite   24
        1.2.4   Miscellaneous   25
    1.3   Artificial Actinides   25
        1.3.1   Actinide production in the nuclear fuel cycle   25
        1.3.2   Weapons-grade plutonium   29
        1.3.3   Minor actinides   33
                 1.3.3.1   Neptunium-237   33
                 1.3.3.2   Americium   34

           1.3.3.3   Curium   35
           1.3.3.4   Berkelium and Californium   36
    1.4   Actinide Host-Phases   36
           1.4.1   Natural accessory minerals   36
           1.4.2   Zircon and hafnon   38
           1.4.3   Monazite   41
           1.4.4   Zirconolite   41
           1.4.5   Baddeleyite (monoclinic zirconia)   42
           1.4.6   Tazheranite (cubic zirconia)   43
           1.4.7   Xenotime   43
           1.4.8   Apatite   44
           1.4.9   Pyrochlore   45
           1.4.10   Perovskite   46
           1.4.11   Garnet   47
           1.4.12   Murataite   49
           1.4.13   Kosnarite   49
           1.4.14   Natural gels   50
    References   52

2. **Current and Potential Actinide Applications**   **57**

    2.1   Advanced Nuclear Fuel Cycle   57
           2.1.1   MOX nuclear fuel   57
           2.1.2   Ceramic nuclear fuel   58
           2.1.3   Advanced nuclear reactors   59
    2.2   Inert Pu Ceramic Fuel   60
    2.3   Sealed Radioactive Sources   61
    2.4   Self-glowing Materials   65
    2.5   Transmutation Targets   68
    2.6   Summary   69
    References   69

3. **Waste Actinide Immobilisation**   **73**

    3.1   Ceramic Nuclear Wasteforms: Historical
           Overview   73
           3.1.1   Early work   73
           3.1.2   Emergence of Pu wasteforms   75

	3.1.3	Emergence of durability studies	78
3.2	Titanate-based Ceramics		83
	3.2.1	Synroc	83
	3.2.2	Ti-pyrochlore	84
3.3	Phosphate-based Ceramics		87
	3.3.1	Monazite	87
	3.3.2	Th-phosphate/diphosphate (TPD)	87
	3.3.3	Kosnarite and NZP	88
	3.3.4	Apatite	88
3.4	Ceramics Based on Zirconium and Hafnium Minerals		89
	3.4.1	Zircon/zirconia and hafnon/hafnia	89
	3.4.2	Cubic zirconia (tazheranite) and hafnia . . .	92
3.5	Garnet/perovskite		94
3.6	Summary		100
References			100

4. Synthesis Methods   111

4.1	Precursor Fabrication		111
	4.1.1	Sol-gel	111
	4.1.2	Co-precipitation	115
	4.1.3	Oxide powder mix	116
4.2	Hot Uniaxial Pressing (HUP)		117
4.3	Hot Isostatic Pressing (HIP)		118
4.4	Pressing-sintering		118
4.5	Melting-crystallisation		120
4.6	Self-sustaining (Self-propagating) High Temperature Reactions		122
4.7	Single Crystal Growth		123
4.8	Summary		128
References			129

5. Examination of Highly Radioactive Samples   133

5.1	XRD Analysis		133
5.2	SEM and EPMA		134
5.3	Cathodoluminescence		139

5.4   Optical Microscopy 142
5.5   Mechanical Durability 143
5.6   Leach and Alteration Tests 145
References   152

6. Radiation Damage   157

6.1   Ion-irradiation   164
6.2   Doping with ^{238}Pu and ^{244}Cm 166
       6.2.1   Zircon/zirconia and
                hafnon/hafnia ceramics 166
       6.2.2   Zircon single crystal 168
       6.2.3   Cubic zirconia ceramic 170
       6.2.4   Monazite ceramic 171
       6.2.5   Monazite single crystal 172
       6.2.6   Ti-pyrochlore ceramic 173
       6.2.7   Zr-pyrochlore ceramic 177
       6.2.8   Zirconolite ceramic 178
       6.2.9   Garnet ceramic 180
       6.2.10   Silicate-apatite ceramic
                and chlorine-apatite powder 180
6.3   Main points from Self-irradiation/radiation
       Damage Studies 181
References   181

7. What is the Future?   189

7.1   Safety Issues 189
7.2   Burning (Transmutation Option) 190
7.3   Disposal of Waste Actinides 190
7.4   Performance of Actinides in Disposal
       Environment 192
7.5   Conclusions 192

Index   195

Chapter 1

Introduction to the Actinides

We first introduce the history of the actinides elements and their basic physical and chemical properties and then give a brief description of natural and artificial actinide-containing crystals. Actinides such as Pu are presently used as sources of energy in nuclear reactors and in fundamental physics and chemistry studies including the search for super-heavy metastable nuclides. However, minor actinides (MA), such as Np, Am and Cm are considered waste and many studies are aimed towards their immobilisation in a suitable host for eventual geological disposal.

1.1. Actinide Series

1.1.1. *History*

Natural actinides were discovered more than 200 years ago (see Table 1.1.1). Natural uranium and thorium found limited use in the 18th and 19th centuries when they were of fundamental interest only. The search for new actinides is ongoing (Table 1.1.1).

In 1896, Henri Becquerel studied the phosphorescence of different salts including $K_2(UO_2)(SO_4)_2$. He was continuing the research started by his father and grandfather. Becquerel believed that after exposure to sunlight some salts emitted not only visible light but also

2   *Crystalline Materials for Actinide Immobilisation*

Table 1.1.1.   Historical and chemical details of actinides.

Element, symbol, atomic number	Year of discovery	Material used for discovery	Content in natural materials	Typical valence state	Electronic structure
Uranium, U, 92	1789	Natural uranium oxide ore (pitchblende)	$(2-4)\ 10^{-4}$ wt.% in rocks; $4\ 10^{-4}$ wt.% in ocean; $(1-6)\ 10^{-6}$ g/l in water streams	6 and 4	$5f^36d7s^2$
Thorium, Th, 90	1828	Natural ThSiO$_4$ (thorite)	$(1-2)\ 10^{-3}$ in rocks	4	$6d^27s^2$
		1896 — Discovery of radioactivity phenomena			
Actinium, Ac, 89	1899–1904	Rare-earth fraction after chemical processing of natural uranium oxide ore (pitchblende)	$1.5\ 10^{-4}$ g ^{227}Ac in 1 tonne of natural uranium oxide ore (pitchblende); $5\ 10^{-8}$ g ^{228}Ac (estimated) in 1 tonne of natural Th	3	$6d7s^2$
Protactinium, Pa, 91	1913–1917	Natural uranium oxide ore (pitchblende)	$0.10-0.34$ g in 1 tonne of natural uranium oxide ore (pitchblende)	5	$5f^26d7s^2$ or $5f^16d^27s^2$
		1919 — First artificial nuclear reaction			
		1932 — Discovery of neutron			
		1934 — Discovery of artificial radioactivity			
		1938 — Discovery of uranium fission under neutron irradiation			
		1940 — Discovery of uranium spontaneous decay			

(Continued)

Table 1.1.   (*Continued*)

Element, symbol, atomic number	Year of discovery	Material used for discovery	Content in natural materials	Typical valence state	Electronic structure
Neptunium, Np, 93 (First artificial actinide)	1939–1940	^{239}Np — from ^{238}U irradiated by neutrons	1.8 10^{-12} parts of ^{237}Np per one part of ^{238}U in natural uranium oxide ore (pitchblende) from Congo	5 and 4	$5f^4 6d^1 7s^2$
Plutonium, Pu, 94	1940	^{238}Pu — from ^{238}U irradiated by 14 MeV deuterons in cyclotron; 1941, ^{239}Pu as a fission product of ^{239}Np	(0.4–15) 10^{-12} parts of ^{239}Pu per part of ^{238}U in natural uranium minerals; 7.1 10^{-12} parts of ^{238}Pu in natural uranium minerals; (0.7–2.0) 10^{-9} of ^{239}Pu in natural uranium oxide ore (pitchblende) and monazite*; (1–7) 10^{-13} g/l ^{239}Pu measured in samples of natural water*; (1.7–2.9) 10^{-10} wt.% ^{239}Pu measured in samples of natural zircon*	4 and 3	$5f^6 7s^2$
Americium, Am, 95	1944–1945	From ^{239}Pu highly irradiated by neutrons	—	3 and 4	$5f^7 7s^2$

(*Continued*)

4   *Crystalline Materials for Actinide Immobilisation*

Table 1.1.1.   (*Continued*)

Element, symbol, atomic number	Year of discovery	Material used for discovery	Content in natural materials	Typical valence state	Electronic structure
Curium, Cm, 96	1944–1945	^{242}Cm — from ^{239}Pu irradiated by 32 MeV alpha-particles	As assumed, only ^{247}Cm might be identified in rare-earth minerals at level 10^{-8} wt.%	3	$5f^7 6d7s^2$
Berkelium, Bk, 97	1950	From ^{242}Cm bombarded by 35 MeV alpha-particles in cyclotron	—	3	$5f^{10}7s^2$
Californium, Cf, 98	1949	^{243}Bk — from ^{241}Am irradiated by 35 MeV alpha-particles	—	3	$5f^8 6d7s^2$ or $5f^9 7s^2$
Einsteinium, Es, 99	1952	Product of thermonuclear explosion "Mike" (first thermonuclear bomb)	—	3	$5f^{11}7s^2$
Fermium, Fm, 100	1952	Product of thermonuclear explosion "Mike" (first thermonuclear bomb)	—	3	$5f^{12}7s^2$
Mendelevium, Md, 101	1955	From ^{253}Es bombarded by 35 MeV alpha-particles in cyclotron			$5f^{13}7s^2$

To be continued with newly discovered elements

*(Cherdintsev *et al.*, 1965)

some unknown rays, which were similar to the X-rays discovered in 1895 by Wilhelm Roentgen. One gloomy day, Becquerel left some uranium salt and a photographic plate in the same box and later he found that the plate had an exposed image. Further experiments excluded any other possible explanation of this observation except from the properties of the uranium salt: Becquerel concluded uranium atoms were the source of the invisible rays.

With the discovery of radioactivity phenomena, interest in uranium and thorium increased dramatically. A large uranium deposit was discovered in Jachymov (now in the Czech Republic). Meanwhile, in Paris, Marie and Pierre Curie began to investigate uranium ores and discovered that natural uranium oxide ore (pitchblende) was more radioactive than the pure uranium extracted from this ore. Further chemical processing of tonnes of uranium by the Curies allowed concentration of trace radioactive elements and led to the discovery of radium (Ra), polonium (Po), actinium (Ac), protactinium (Pa) and radon (Rn), for example. At this time radium was popular among scientists examining the structure of atomic nuclei. For decades, uranium ores were used as the main source of radium. Actinium and protactinium were also extracted in small amounts from uranium ores.

In 1919, at the University of Manchester, England, Ernest Rutherford created the first man-made nuclear reaction through bombardment of nitrogen nuclei by alpha-particles. As a result, reaction nitrogen was converted into an isotope of oxygen.

Much progress in the nuclear field was made during the 1930s and 40s. In 1932, James Chadwick found that irradiation of beryllium by alpha-particles caused the emission of "neutral particles", which were known as "neutrons". Then, in 1934, Frederic and Irene Joliot-Curie discovered the phenomena of artificial radioactivity and observed that bombardment of boron, aluminium and magnesium with alpha-particles caused nuclear reactions and formation of new radioactive elements, such as nitrogen, phosphorus and silicon, respectively. Targets bombarded by alpha-particles emitted neutrons. In 1944, Enrico Fermi in Rome suggested using neutron bombardment to synthesise new radioactive elements. At that time, the only available source of neutrons was a mixture of radium or polonium with beryllium. Fermi

studied different elements irradiated by neutrons and assumed uranium that had captured a neutron must be converted into heavier
beta-radioactive uranium isotope. Decay of this isotope had to give
birth to new transuranium element #93. All other attempts to identify new transuranium elements at this time by studying uranium
irradiated by neutrons were unsuccessful, however.

In Berlin, Otto Hahn and Fritz Strassmann observed fission of
uranium nuclei under neutron irradiation in 1939. Measurements of
the energy released during this reaction were performed worldwide by
various research groups during World War II. In fact, it was surprisingly high — approximately 200 MeV. Frederic and Irene Joliot-Curie
had already proven that decay of each uranium nucleus is accompanied by release of more than 2 neutrons. On 2 August 1939, Albert
Einstein sent a letter to US President Franklin Delano Roosevelt
in which he drew attention to nuclear research related to uranium,
which could lead to the construction of extremely powerful new types
of bomb.

In 1940, in Leningrad (now St Petersburg), Russia, Georgii Flerov
and Konstantin Petrzhak (1940) discovered spontaneous (without
neutron participation) fission of uranium, accompanied by the emission of several neutrons and $\sim$200 MeV energy release for each fission
event. To verify and confirm this discovery, they repeated the experiment underground in Moscow's Metro, where conditions provided
protection against cosmic radiation. The probability of spontaneous
fission is very low (several atoms decay each hour in 1 g of uranium), but even so they perceived it might give enough neutrons
to initiate a nuclear chain reaction in uranium. At about this time
Edwin McMillan and Philip Abelson (Berkeley Radiation Laboratory, University of California) discovered ^{239}Np as a result of ^{239}U
decay formed through neutron irradiation of ^{238}U. "Neptunium" was
named after the planet Neptune, which is located in the Solar System
behind Uranus (after which U had been named).

In 1941, Glenn Seaborg, Edwin McMillan, Joseph Kennedy and
Arthur Wahl (Berkeley, California) discovered ^{238}Pu after deuteron
bombardment of ^{238}U in a cyclotron. This element was named "plutonium" after the planet Pluto located after Neptune in the Solar

System. Later, neutron irradiation of ^{238}U led to ^{239}Pu (Seaborg *et al.*, 1946). It was clear that the nuclear properties of ^{239}Pu which might support initiation of a chain nuclear reaction had to be similar to ^{235}U. However, chemical extraction of plutonium is essentially easier than separation of ^{235}U and ^{238}U. By December 1943, only 2 mg plutonium was available in the USA (Groves, 1964). Indeed, all information about plutonium was classified until 1946.

In 1944–1945, Glenn Seaborg, with colleagues Ralf James and Leon Morgan, discovered americium and then, with James and Albert Ghiorso, curium. Both elements were obtained as a result of ^{239}Pu irradiation — by neutrons (^{241}Am) or alpha-particles (^{242}Cm). Americium was named after "America" in a similar way to its lanthanide analogue europium (after "Europe"). The name "curium" was dedicated to Marie and Pierre Curie.

At the start of 1945 in the Radium Institute, Leningrad, USSR (now the V.G. Khlopin Radium Institute (KRI), St Petersburg) the first sample of plutonium was obtained by neutron irradiation of uranium in a cyclotron (Fig. 1.1.1). The amount of ^{239}Pu was extremely low and it was detected by an alpha-counter as 33 events per minute (KRI-75, 1997). The first Soviet technology on plutonium

Fig. 1.1.1.   The first European cyclotron (magnet with accelerating camera) built in 1937 at the Radium Institute (Leningrad, USSR) and used to obtain the first Soviet sample of plutonium in 1945.

extraction and purification was developed using approximately 1 μg of plutonium.

On the 16th July 1945, in New Mexico, USA, the first nuclear explosion of a plutonium device (6.1 kg of Pu) was carried out, and on 6th August 1945, the nuclear bombardment of the city of Hiroshima, Japan, by the USA was the first explosion of a uranium atomic bomb (60 kg of highly enriched U). The first plutonium atomic bomb (6.1 kg of Pu) was dropped on Nagasaki, Japan on 9th August 1945.

Following this, the first Soviet plutonium atomic bomb exploded on 29th August 1949 in the Semipalatinsk region (now the Kazakhstan Republic), while the first British nuclear bomb was tested on 3rd October 1952 on the Montebello Islands in Western Australia. The bomb used plutonium produced mainly at Windscale and was a close copy of the US bomb used at Nagasaki.

1.1.2. *Basic physical and chemical properties*

The principal features of actinides are summarised in Tables 1.1.2–1.1.4. Plutonium is the only chemical element with six allotropes in the metallic state. Uranium is characterised by many oxidation states and forms many oxide polymorphs. There are some similarities in chemical behaviour between actinides and lanthanides: however, there are no ideal chemical analogues (or simulants) among lanthanides for Th, Pa, U, Pu and Np. From U to Am the orbital energy for $5f$-, $6d$-, $7s$- and $7p$- is approximately the same (Cotton and Wilkinson, 1988). This gives rise to far greater chemical diversity in the actinides than is observed in the lanthanides.

1.1.3. *History of using actinide-containing materials*

The two main natural actinides, uranium and thorium, were the first actinide elements to find practical use. From the 18th century, uranium salts and oxides were well known in Europe as glass and porcelain colorants conferring green and yellow. However, uranium-bearing glass (with a light-green colour) is found in ancient Roman mosaics. Native Indians in America used uranium for colouring glazed ceramic. Thorium, which does not form simple coloured chemicals, was not

Table 1.1.2.   Basic features of actinide isotopes (Yagovkina, 2009).

Element	Isotope	Half-life (years)	Activity per gramme Bq	Activity per gramme Ci
Actinium	^{227}Ac	21.8	$2.7 \ 10^{12}$	72
Thorium	^{228}Th	1.9	$3.0 \ 10^{13}$	820
	^{229}Th	7,340	$7.9 \ 10^{9}$	0.2
	^{230}Th	$7.70 \ 10^{4}$	$7.5 \ 10^{8}$	0.02
	^{232}Th	$1.41 \ 10^{10}$	$4.1 \ 10^{3}$	$1.1 \ 10^{-07}$
Protactinium	^{231}Pa	$3.28 \ 10^{4}$	$1.7 \ 10^{9}$	0.05
Uranium	^{232}U	68.9	$8.3 \ 10^{11}$	22
	^{233}U	$1.59 \ 10^{5}$	$3.6 \ 10^{8}$	0.01
	^{234}U	$2.45 \ 10^{5}$	$2.3 \ 10^{8}$	0.006
	^{235}U	$7.04 \ 10^{8}$	$8.0 \ 10^{4}$	$2.2 \ 10^{-6}$
	^{236}U	$2.34 \ 10^{7}$	$2.4 \ 10^{6}$	$6.5 \ 10^{-5}$
	^{238}U	$4.47 \ 10^{9}$	$1.2 \ 10^{4}$	$3.4 \ 10^{-7}$
Neptunium	^{237}Np	$2.14 \ 10^{6}$	$2.6 \ 10^{7}$	$7.0 \ 10^{-4}$
Plutonium	^{235}Pu	$8.11 \ 10^{-7}$	$6.9 \ 10^{19}$	$1.9 \ 10^{9}$
	^{236}Pu	2.9	$2.0 \ 10^{13}$	530
	^{238}Pu	87.7	6.3E+11	17
	^{239}Pu	$2.41 \ 10^{4}$	$2.3 \ 10^{9}$	0.06
	^{240}Pu	6,570	$8.4 \ 10^{9}$	0.2
	^{241}Pu	14.4	$3.8 \ 10^{12}$	100
	^{242}Pu	$3.76 \ 10^{5}$	$1.5 \ 10^{8}$	0.004
Americium	^{241}Am	432.1	$1.3 \ 10^{11}$	3.4
	^{242}Am	152	$3.6 \ 10^{11}$	9.7
	^{243}Am	7,380	$7.4 \ 10^{9}$	0.2
Curium	^{242}Cm	0.45	$1.2 \ 10^{14}$	3,300
	^{244}Cm	18.1	$3.0 \ 10^{12}$	81
	^{245}Cm	8,500	$6.4 \ 10^{9}$	0.2
	^{246}Cm	4,730	$1.1 \ 10^{10}$	0.3
	^{247}Cm	$1.56 \ 10^{7}$	$3.4 \ 10^{6}$	$9.1 \ 10^{-5}$
	^{248}Cm	$3.39 \ 10^{5}$	$1.6 \ 10^{8}$	0.004
Californium	^{251}Cf	898	$5.7 \ 10^{10}$	1.6
	^{252}Cf	2.6	$2.0 \ 10^{13}$	540

Table 1.1.3.   Crystal structures of actinide metals (Yagovkina, 2009).

Actinide	Polymorph	Density, g/cm^3	Crystalline structure	Unit cell parameters, Å
Ac		10.066	Face-centred cubic	a = 5.311
Th	Alpha (α)		Face-centred cubic	a = 5.0722
	Beta (β)		Body-centred cubic	a = 4.11
Pa		12.051	Body-centred cubic	a = 5.031
		15.382	Body-centred tetragonal	a = 3.925
				c = 3.238
U		18.660	Simple tetragonal	a = 10.61
				c = 5.645
	Alpha (α)	19.067	Body-centred orthorhombic	a = 2.8548
				b = 5.8589
				c = 4.9576
Np	Alpha (α)	20.487	Simple orthorhombic	a = 4.721
				b = 4.888
				c = 6.661
	Beta (β)	19.379	Simple tetragonal	a = 4.897
				c = 3.388
Pu	Alpha (α)	20.262 (25°C)	Simple monoclinic	a = 6.183
				b = 4.822
				c = 10.963
				β = 101.8
	Beta (β)	18.28 (150°C)	Body-centred monoclinic	a = 9.227
				b = 10.449
				c = 7.824
				β = 92.54
	Gamma (γ)	17.506 (210°C)	Face-centred orthorhombic	a = 3.1587
				b = 5.7682
				c = 10.162

(*Continued*)

Table 1.1.3.   (*Continued*)

Actinide	Polymorph	Density, g/cm^3	Crystalline structure	Unit cell parameters, Å
	Delta (δ)	16.254 (320°C)	Face-centred cubic	a = 4.6371
	Delta prime (δ')	16.346 (465°C)	Body-centred tetrag-onal	a = 3.339
				c = 4.446
	Epsilon (ε)	16.856 (500°C)	Body-centred cubic	a = 3.6361
Am		11.948	Simple hexagonal	a = 3.642
				c = 11.78
		13.770	Face-centred cubic	a = 4.894
		15.991	Face-centred cubic	a = 4.565
Cm		13.692	Face-centred cubic	a = 4.93
		19.498	Face-centred cubic	a = 4.382
		13.646	Simple hexagonal	a = 3.502
				c = 11.32
Cf		10.170	Face-centred cubic	a = 5.473
		8.744	Simple hexagonal	a = 3.998
				c = 6.887
		15.284	Simple hexagonal	a = 3.38
				c = 11.025
		24.510	Triclinic	a = 3.307
				b = 7.412
				c = 2.793
				α = 89.06
				β = 85.15
				γ = 85.7

used until much later. In 1891, Carl Auer von Welsbach found that thorium oxide doped with Ce emits an intensive glow in the flame of an oil or kerosene lamp. This led to a demand for thorium nitrate from lamp manufacturers. Special grids or wicks impregnated with ThO_2 (doped with about 1 wt.% CeO_2) are still used for some camping lamps.

Interestingly, thorium has other lighting applications but in electric incandescent and gas-discharge lamps. Small admixtures of thorium in the tungsten wires of electric lamps increase their life by protecting the tungsten from recrystallisation and delaying creep.

12   *Crystalline Materials for Actinide Immobilisation*

Table 1.1.4.   Crystal structures of actinide oxides (Yagovkina, 2009).

Actinide	Actinide oxide	Density, g/cm^3	Colour	Crystalline structure	Unit cell parameters, Å
Ac	Ac_2O_3	9.19	White	Simple hexagonal	a = 4.078
					c = 6.39
Th	ThO_2	10.00	White	Face-centred cubic fluorite-type	a = 5.597
	ThO	?	Black	Face-centred cubic	a = 5.302
Pa	Pa_2O_5	?	White	Face-centred cubic fluorite-type	a = 5.455
	PaO	13.439	?	Face-centred cubic	a = 4.961
	PaO_2	10.472	Black	Face-centred cubic	a = 5.505
U	UO_2	10.977	Dark-brown	Face-centred cubic fluorite-type	a = 5.467
	α-U_3O_8	12.547	Green-grey	Body-centred orthorhombic	a = 7.062
					b = 3.81
					c = 4.142
	U_3O_7	14.537	?	Tetragonal	a = 5.3811
					c = 5.5400
	U_2O_5	8.351	?	Orthorhombic	a = 8.29
					b = 31.71
					c = 6.73
	α-UO_3	8.3	Brown	Hexagonal	a = 3.971
					c = 4.27
	α-UO_3	8.017	?	Monoclinic	a = 6.895
					b = 19.94
					c = 6.895
					β = 90.4
	β-UO_3	8.0	Orange	Monoclinic	a = 10.34
					b = 14.33
					c = 3.910
					β = ?

(*Continued*)

Table 1.1.4.   (*Continued*)

Actinide	Actinide oxide	Density, g/cm^3	Colour	Crystalline structure	Unit cell parameters, Å
	γ-UO$_3$	7.3	Yellow	Tetragonal	a = 6.89 c = 19.94
	δ-UO$_3$	6.7	Dark-red	Cubic	a = 4.16
	UO$_3$	8.7	Red	Triclinic	a = 4.002 b = 3.841 c = 4.165
Np	NpO$_2$	11.10	Brown-black	Face-centred cubic fluorite-type	a = 5.433(1) – single crystals synthesised at KRI
Pu	PuO$_2$	11.662	Yellow, yellow-green, khaki, dark-brown. Black (single crystals)	Face-centred cubic fluorite-type	a = 5.397
					a = 5.397(1) – single crystals synthesised at KRI
	PuO	14.170	Black	Face-centred cubic fluorite-type	a = 4.958
	α-Pu$_2$O$_3$	10.670	?	Body-centred cubic	a = 11.01
	β-Pu$_2$O$_3$	11.789	?	Hexagonal	a = 3.838 c = 5.918
Am	AmO$_2$	11.678	Black	Face-centred cubic fluorite-type	a = 5.388
	AmO	13.397	?	Face-centred cubic	a = 5.045
	Am$_2$O$_3$	?	Brown	Hexagonal	a = 3.817 c = 5.971
	Am$_2$O$_3$	10.573	Red-brown	Body-centred cubic	a = 11.03
	CmO$_2$	11.994	Black	Face-centred cubic fluorite-type	a = 5.366

(*Continued*)

14   *Crystalline Materials for Actinide Immobilisation*

Table 1.1.4.   (*Continued*)

Actinide	Actinide oxide	Density, g/cm³	Colour	Crystalline structure	Unit cell parameters, Å
Cm	Cm_2O_3	10.831	White	Body-centred cubic	a = 5.50
	Cm_2O_3	12.019	?	Hexagonal	a = 3.799
					c = 5.991
	Cm_2O_3	11.823	?	Body-centred monoclinic	a = 14.282
					b = 3.652
					c = 8.9
					β = 100.3
	Cm_2O_{12}	11.261	?	Rhombohedral	a = 10.19
					c = 9.45
Cf	Cf_2O_3	12.466	?	Body-centred monoclinic	a = 14.124
					b = 3.591
					c = 8.809
	CfO_2	12.541	?	Face-centred cubic	a = 5.312
	Cf_2O_3	11.478	?	Face-centred cubic	a = 10.838
	Cf_2O_3	12.786	?	Hexagonal	a = 3.72
					c = 5.96

In addition to thorium's suitable chemical properties, its alpha-radiation (but the low radiotoxicity of ^{232}Th) is also applied in powerful modern gas-discharge lamps (up to 3,500 W) based on the fluorescence of Hg-halides (Fig. 1.1.2). Use of small amounts of Th (10^{-4} g per lamp) encourages electric-arc formation when the lamp is turned on. Admixtures of Th-iodide in Hg-halides are used to change the spectral features of the light emission.

Thorium dioxide is a chemically inert refractory material with melting point 3,050°C, which led to its use in refractory crucibles for high-temperature experiments in air (other refractory crucible materials include ZrO_2 and MgO, which melt at 2,700 and 2,825°C, respectively).

Since the end of the 19th century natural uranium ores have been the source of radium, which initially attracted the attention of the

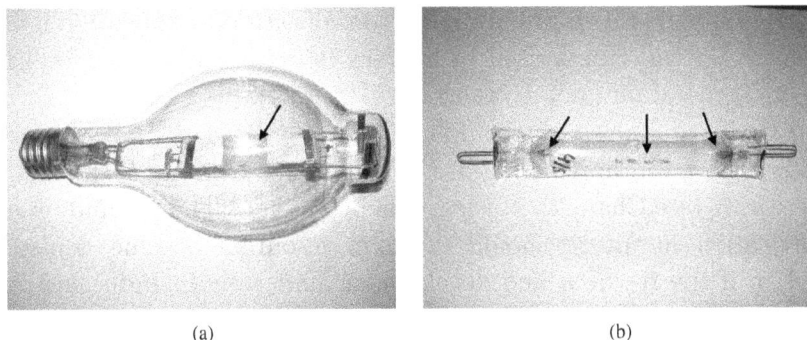

(a)   (b)

Fig. 1.1.2.   (a) Powerful gas-discharge lamp using Th-doped tungsten electrodes and dispersed Th inside sealed silica-glass capsule; (b) the initial state of lamp-fuse before activation and fabrication of the whole lamp. Pellets of Hg-salts mixed with powder of metallic Th (central arrow) are placed into a glass chamber filled with argon. First, strong electric discharge between two Th-doped tungsten electrodes (left and right arrows, respectively) evaporates Hg-salts and Th. This activates the lamp-fuse for further use. Ionisation of inert gas by alpha-radiation from Th stabilises electric-arc initiation when the lamp is turned on.

world's nuclear research community. Later, uranium became important as an element of the first atomic bombs. Although ^{235}U is still a weapon material, its use in modern nuclear weapons is limited and it has been largely replaced in this application by ^{239}Pu. However, ^{233}U as a transmutation product of thorium irradiation may compete with ^{239}Pu in weapon applications if the thorium nuclear cycle is used in future. Metallic ^{238}U is sometimes used for protection against hard gamma-radiation, but at the same time it is used in thermo-nuclear weapons. The pyrophoric nature of uranium, along with its high density, meant it found application in "regular" weapons. Special armour-piercing shells and bombs with pits and cores made of depleted metallic uranium (DU) are used to penetrate the armour of tanks and other military vehicles. Such "regular" weapons were widely used by US forces during the Iraq War in 1991 and by NATO forces throughout the Balkan War (1999). According to some sources (Guardian, 2000) 10 tonnes of depleted uranium were used in military operations alone in Kosovo, causing soil contamination at dozens of sites. Some military training sites (even in the UK at Eskmeals, Cumbria) are extensively contaminated with DU.

16   *Crystalline Materials for Actinide Immobilisation*

To complete the military applications of actinides we have to mention ^{227}Ac, which was used in igniters (special neutron sources) in second-generation nuclear weapons.

"Civilian" nuclear fuel applications of ^{238}U and ^{235}U are well known, however, application of ^{239}Pu in mixed oxides (MOX) fuel is increasing (see Chap. 2). Nuclear fuel based on ^{232}Th may find broad application in future should uranium resources become depleted and/or if the research and development underway in India and the USA is successful.

The actinide isotopes characterised essentially by heat emission (^{238}Pu, ^{242}Cm, ^{244}Cm) have potential applications in electric-heat generators. US space vehicles use ^{238}Pu-based sources of electricity with power outputs from 2.9 to 25 W. In 1978, a Soviet ^{238}Pu-based heat source "Vysota" ("Height"), with 1,000 W power and operating temperature 950°C, was developed (KRI-75, 1997). Very pure ^{238}Pu (see Table 1.3.1) in the form of PuO_2 has been suggested in Russia and other countries for use in heart pacemakers. However, natural oxygen consists of three isotopes (in wt.%): ^{16}O −99.76, ^{17}O −0.04, ^{18}O −0.20 and is not optimal due to high neutron emission from alpha-irradiated ^{17}O and ^{18}O. 1 g of ^{238}Pu emits only 2,600 neutrons per second as a result of spontaneous fission. In oxide form (with natural oxygen) neutron emission from 1 g of ^{238}Pu exceeds health limits ($1.7 \cdot 10^4$ neutrons/sec). Therefore, production of ^{238}Pu for medical purposes requires depletion of ^{17}O and ^{18}O to a total content of both isotopes up to 0.01 wt.% (instead of 0.24 wt.%). Neutron emission from this "improved" ^{238}PuO_2, developed at the V.G. Khlopin Radium Institute and produced in macro-quantities at Russia's Mayak radiochemical plant was acceptable for medical use ($1.68 \cdot 10^4$ neutrons/sec for 1 g of ^{238}Pu), but it has not been used in practice (KRI-75, 1997).

Some actinide compounds reveal superconductivity, e.g., UPt$_3$ (critical temperature, $T_c = 0.48$ K), UBe$_{13}$ ($T_c = 0.85$ K), PuCoGa$_5$ ($T_c = 18.5$ K) and PuRhGa$_5$ ($T_c = 9$ K) but future application of these superconductors appears unlikely since inactive superconductors with higher T_cs are available.

Some actinides are used as neutron sources of two types: so-called "alpha-neutron" or "α/n" sources based on ^{227}Ac or ^{238}Pu and Be (which emits neutrons under alpha irradiation) and "direct" sources such as ^{252}Cf and ^{254}Cf based on neutron emission as a result of spontaneous fission (1 g of ^{252}Cf and 1 g of ^{254}Cf give $3.03 \cdot 10^{12}$ n/sec and $1.35 \cdot 10^{15}$ n/sec, respectively). However, handling of ^{252}Cf and ^{254}Cf requires hot cells with neutron-absorbing concrete and thick windows filled with a high-density aqueous solution of $ZnBr_2$ as a radiation shield.

Americium is used extensively in smoke detectors (in quantities of a few tenths of a μg of ^{241}Am).

In summary, the actinide family is characterised by properties that are attractive and promising for broad application in many fields of human endeavour. However, there are strict barriers that currently prevent their application, including the threat of nuclear proliferation, the high level of radioactivity of most actinides and serious concerns related to their radiotoxicity.

1.1.4. *High toxicity and long-lived radioactivity*

Most actinide isotopes, with the exception of ^{232}Th, ^{238}U and ^{235}U, are extremely dangerous and their internal uptake during professional handling is strictly limited (see Table 1.1.2). The main hazard associated with actinides is related to alpha-radiation although some actinides are characterised by gamma and neutron emission too (particularly in mixtures with Be and other light elements) and they may also be chemically toxic. For example, the lethal ^{235}U content in animals is only 2 mg/kg.

Alpha-particles do not travel long distances. Each one moving in air creates $(1.2–3.5) \cdot 10^5$ ion pairs along its path. Their range of travel does not exceed several centimetres in dry air and only dozens of micrometres in biological tissue. High density precludes the alpha particles from penetrating most solid materials. Therefore, a piece of paper or human skin provides adequate shielding against alpha-radiation. However, inhaled alpha-radioactive compounds cause significant damage to biological tissue through breaking bonds in organic molecules and formation of carcinogenic-free radicals. 5 MeV alpha-particles travel for approximately 40 μm in biological tissue and can penetrate more than 10,000 cells within their range. At pH values found in the human body most actinides can form radio-colloids in liquids, which might overcome more effectively the biological barriers in comparison with ionic forms of actinides. At the same time radio-colloids may be absorbed by biological tissue and they are difficult to excrete. About 10% of inhaled Pu and 25% of inhaled U remain in the human body. The time of the Pu biological cycle (before complete release from the body due to biochemical exchange) is approximately 100 years.

Extremely low contents of highly alpha-radioactive actinides penetrating a mammal body can cause cancer. It has been reported (Berdjis, 1971) that lung cancer occurred in dogs that inhaled $^{239}PuO_2$ at level of μg per kilogram body weight, while bone cancer occurred by doses of 0.26 μg/kg. A statistically significant excess of lymphopoiesis tumours was observed in Rocky Flats (US) employees having internal ^{239}Pu contents greater than 0.03 μg. Long-term chronic toxicity of most long-lived actinides is of much greater concern than their immediate acute toxicity. For example, for Pu isotopes a latent period of about 20 years is expected between exposure and potential development of a solid tumour in the lung or bone (Hoffman, 2002). In the human body soluble forms of actinides, which enter the blood stream from the lungs or from the stomach-bowel system, are distributed among bone, kidneys and liver. Accumulation of actinides in a particular human organ depends on the actinides' chemical form and valence state. For example, Pu(3+), Np(5+) and Np(6+) are absorbed mainly in near-surface bone layers, but Pu(4+) and Np(4+) accumulate preferentially in liver tissue. Actinides accumulated in the liver over a long time may move into bone. Inhalation of different forms of plutonium dioxide results in a range of biological consequences (ICRP, 1986). $^{239}PuO_2$ produced at temperatures greater than 600°C does not leave the lungs after inhalation. On the contrary, approximately 20% of $^{239}PuO_2$, which was prepared at room temperature, goes from the lungs to the blood and then accumulates in the skeleton and liver. Inhaled $^{238}PuO_2$ is transported from the respiratory to the blood system at least three times faster than low-temperature $^{239}PuO_2$ (Il'in and Filatov, 1990).

Aqueous solutions of actinide salts may penetrate the skin. After five minutes' exposure to 6.7% aqueous solution of uranyl-nitrate on skin, concentrations of uranium in the blood as high as 0.2–1.0 μg/ml have been observed. Incorporation of Pu and Am from solutions through the skin might vary in a broad range from 10^{-2} to several % and increase by 100–250 times for damaged skin (Il'in and Filatov, 1990).

For those living around radiochemical plants or working close to nuclear related processes (but not working in them), there are strict limitations concerning the content of the main isotopes of Pu, Np, Am and Cm: around 10^{-6} Bq/l in air and 70–130 Bq/l in water.

All actinides are capable of being highly mobile in the environment and this is a serious concern given the long lives of isotopes such as ^{237}Np, ^{239}Pu and ^{241}Am (see Table 1.1.2). Unsealed powders of oxides of strong alpha-emitters form self-spreading radioactive dust and aerosols. As a result of radiolysis, concentrated aqueous solutions of actinides generate H_2O_2, which may be released via radioactive aerosols.

It is worth noting the current trend to decrease the minimum levels of actinides which require state licensing for any kind of professional handling. This tendency is not based on new data of actinide radiotoxicity but mainly reflects official efforts to minimise the risk of actinides being acquired and used in terrorist attacks. For example, the minimum amount of ^{232}Th that requires licensing in Russia was recently decreased from 900 g (see Table 1.1.5) to 0.25 g (Standards of Radiation Safety, 1999).

1.1.5. *Need for actinide immobilisation*

Most actinide isotopes are extremely dangerous and they require immobilisation. In general, immobilisation means radionuclide conversion into an environmentally safe state (Ojovan and Lee, 2005). However, there are some variations in what is meant by immobilisation. The International Atomic Energy Agency (IAEA) defines immobilisation as the conversion of waste into a wasteform

20   *Crystalline Materials for Actinide Immobilisation*

Table 1.1.5.   Russian state safety standard requirements for work with actinides (Standards of Radiation Safety NRB-76/87, 1988).

Isotope	Annual uptake limit through lungs for those involved in nuclear work (μg)			Minimal activity of open source (amount of actinide) at place of work, for which licensing is required μCi (μg)
	Bone	Kidneys	Lungs	
^{227}Ac	0.00004	0.0004	0.0004	0.1 (0.001)
^{230}Th	0.1	0.3	0.6	0.1 (5)
^{232}Th	20,000	70,000	100,000	100 (900,000,000)
^{231}Pa	0.03	—	3	0.1 (2)
^{233}U	150	70	15	1 (100)
^{234}U	240	110	—	1 (160)
^{235}U	370,000	300,000	80,000	1 (50,000)
^{236}U	10,000	20,000	2,000	1 (20,000)
^{238}U	2,300,000	550,000	480,000	100 (300,000,000)
^{237}Np	—	7	210	0.1 (140)
^{238}Pu	0.0001	—	0.003	0.1 (0.006)
^{239}Pu	0.03	—	0.8	0.1 (1.6)
^{240}Pu	0.009	—	0.2	0.1 (0.4)
^{242}Pu	0.6	—	15	0.1 (26)
^{241}Am	0.002	0.002	0.04	0.1 (0.03)
^{243}Am	0.04	0.04	65	0.1 (0.5)
^{244}Cm	0.0002	—	0.002	0.1 (0.001)
^{245}Cm	0.004	—	0.8	0.1 (0.6)
^{246}Cm	0.02	—	0.4	0.1 (0.3)
^{247}Cm	67	—	1444	0.1 (1100)
^{248}Cm	0.2	—	4	0.1 (20)
^{251}Cf	0.001	—	0.08	0.1 (0.06)

by solidification, embedding or encapsulation with the aim of reducing the potential for migration or dispersion of radionuclides during handling, transport, storage and/or disposal (Radioactive Waste Management Glossary, 2003). We insist on a stricter definition of actinide immobilisation taking into account their potential practical use as well as their eventual disposal. Therefore, *actinide immobilisation is their conversion into the most stable forms to be used as advanced nuclear fuel, special sources of alpha-irradiation, targets for transmutation, durable self-glowing materials, as well as the final wasteforms for geological disposal.*

Immobilisation should achieve the following:

(1) Decreased radiotoxicity danger in actinide handling, transport, storage and disposal due to synthesis of actinide hosts which are not soluble or volatile in different media, including nuclear reactors, geological environment, seawater, space and biological tissue. In most cases mechanical durability and radiation resistance of the actinide hosts are also required to avoid possible dispersion of highly alpha-radioactive materials as dust or "hot particles";

(2) Total exclusion of any use of immobilised actinides directly as nuclear weapon materials or in "dirty bombs" for terrorist attacks.

It is obviously desirable to have immobilised actinides in a form that is difficult to steal and traffic illegally. From 1993–2003, the IAEA reported ten cases of illegal trafficking of highly enriched uranium (20 and more wt.% ^{235}U) and seven cases of plutonium (Orlov, 2004). In one case, 0.3 kg Pu was involved.

1.2. Natural Actinides and Minerals

1.2.1. *Uraninite, pitchblende and thorianite*

Uranium and thorium are the only natural actinides, which themselves form their own (intrinsic) minerals. The geochemistry of uranium and thorium is essentially different and there are no "coupled" U-Th-deposits containing simultaneously intrinsic uranium and thorium minerals, although these elements may accompany each other in the form of chemical admixtures. Natural oxide minerals: uraninite, UO_2, and thorianite, ThO_2, have the same cubic fluorite-type crystalline structure (see Table 1.1.4) and theoretically may form a complete solid solution $(U,Th)O_2$. Not all members of this solid solution series, however, are found in nature. Pure crystalline uraninite (Th-free) and thorianite (U-free) are, however, very rare, being occasionally found in skarns and some hydrothermal rocks formed at high temperature.

Usually, uraninite contains 3–10 wt.% ThO_2, but there are some
examples containing 20–25 wt.% ThO_2. In general, uraninite formed
at higher temperature has higher Th content. Admixture of uranium in thorianite is also typical at a level of 4–14 wt.% UO_2. Thorianite containing 35–38 wt.% UO_2 is called "uranium-thorianite".
Very often uraninite and thorianite contain admixtures of rare-earths
(from several to 10 wt.%). Aged minerals contain radiogenic lead (up
to 10–17 wt.% PbO) and easily detectable radiogenic He (more typical for thorianite). As a result of uranium oxidation, natural uraninite never consists of stoichiometric UO_2, but is better considered as
a mixture of UO_2 and UO_3. Total content of uranium oxide (UO_2
+ UO_3) in uraninite may vary from 67 to 93 wt.% (Soboleva and
Pudovkina, 1957).

Although some natural uraninite samples are millions of years
old, it is not a chemically durable mineral. Under natural conditions
uraninite is substituted by secondary uranium minerals such as uranium silicates (Fig. 1.2.1), or in the case of contact with water and air
by uranium hydroxides and hydroxycarbonates. Considering natural

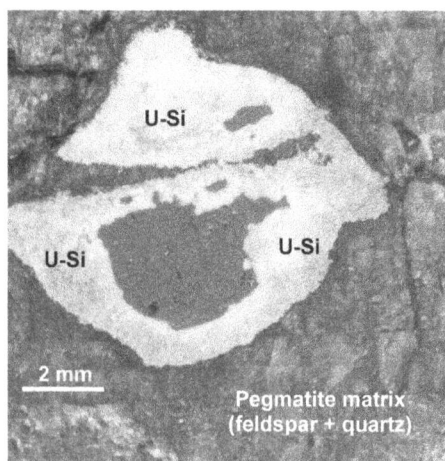

Fig. 1.2.1.   Altered uraninite crystal in pegmatite (Karelia, Russia). Reflected
light (dark field) image of ground (not polished) surface. Dark phases are relicts of
uraninite surrounded by grey-yellow phases (U-Si) of secondary uranium minerals
(U-silicates).

Fig. 1.2.2.   Three cubic crystals of thorianite (centre) among grains of elemental platinum from alluvial placer (Ural region, Russia).

uraninite to be a stable analogue of spent nuclear fuel is incorrect. Natural thorianite, however, is a chemically durable mineral and can be found in alluvial deposits (Fig. 1.2.2).

Another well-known uranium oxide mineral is pitchblende. It is assumed to form from a gel at lower temperatures ($\leq$200–250°C) than uraninite ($\geq$500°C). Its low formation temperature leads to a globular or spherical shape (Fig. 1.2.3). The U oxidation state (ratio UO_2/UO_3) may vary over a broad range in different layers of the same pitchblende globule. Also, in contrast to uraninite, pitchblende is usually free of Th admixture. The world's largest uranium deposits are associated with pitchblende minerals. Natural uranium ore deposits demonstrate the low chemical resistance of pitchblende under environmental conditions since formation of secondary uranium minerals (uranium hydroxides and hydrocarbonates) is observed on the surface of pitchblende ores in the first few months after their release from the host-rock.

1.2.2. *Coffinite and thorite*

The uranium silicate coffinite has the stoichiometric formula $USiO_4$ with a zircon-type tetragonal structure. However, natural coffinite specimens are rare and usually metamict (amorphous, as indicated by

24   *Crystalline Materials for Actinide Immobilisation*

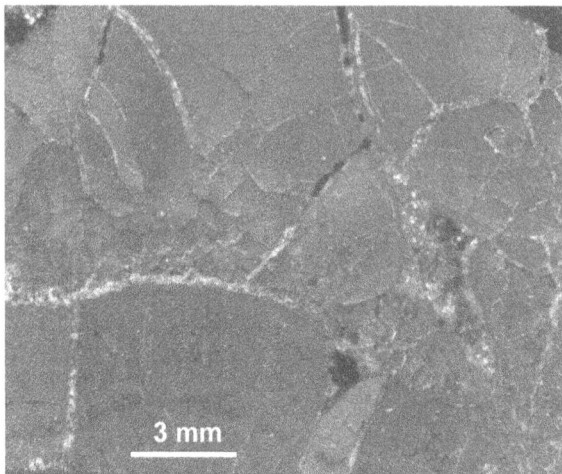

Fig. 1.2.3. Pitchblende (Czech Republic). Reflected light (dark field) image of ground (not polished) surface. White phase is calcite ($CaCO_3$) with grains of pyrite (Fe_2S).

XRD, X-ray diffraction) and may be hydrated. Obtaining synthetic coffinite is difficult because of unclear synthesis conditions.

Thorite, $ThSiO_4$, is isostructural with coffinite but is a more durable mineral. It may be found in alluvial deposits often completely metamict and partly hydrated. Natural thorite may contain 1–2 wt.% UO_2/UO_3, ZrO_2, P_2O_5 and detectable amounts of radiogenic Pb and He. Synthetic thorite may be obtained by a flux method (see Sec. 4.7).

1.2.3. *Brannerite*

Brannerite is uranium titanite with stoichiometric formula UTi_2O_6 and monoclinic crystal structure. Natural brannerite may contain several wt.% CaO and Fe_2O_3, up to 8 wt.% rare earths and ThO_2 (Soboleva and Pudovkina, 1957). It is usually metamict and altered with formation of Ti-hydroxide layers along grain boundaries or cracks inside the crystal matrix. Brannerite is very rarely found in alluvial deposits — however, its chemical durability over long periods is not high.

A natural Th-analogue of brannerite is thorutite, $ThTi_2O_6$. This mineral is extremely rare and its geochemical stability over long times is unclear.

1.2.4. *Miscellaneous*

Study of natural intrinsic uranium minerals (oxides, silicates and titanates) does not support the view that they are chemically and mechanically durable materials that could have "survived" over long periods of time in an open geochemical system. All ancient uranium minerals (up to hundreds of millions of years old) have only been discovered in special host rock environments. Natural weathering of host rocks is accompanied by rapid chemical alteration of all intrinsic uranium minerals.

Natural oxide and silicate minerals of thorium are generally more durable than uranium ones. Despite intrinsic thorium minerals also being affected by radiation damage (metamictisation), hydration and other chemical alteration processes, they have demonstrated long-term resistance to the conditions present in alluvial deposits.

1.3. Artificial Actinides

1.3.1. *Actinide production in the nuclear fuel cycle*

Since the start of the nuclear age the nuclear fuel cycle has had to support production of nuclear weapons. The first two nuclear countries, USA and Russia, chose to develop a uranium fuel cycle to enable production of two main "weapon" isotopes: ^{235}U (from natural ores) and artificial ^{239}Pu (by neutron irradiation of natural ^{238}U). Further civilian use of nuclear energy is based on the uranium fuel cycle as a legacy of the "cold" war and arms race. However, natural thorium might also be used in the nuclear fuel cycle as well as for producing ^{233}U, which is the third important isotope for making nuclear weapons. The simplified schedule of thorium transmutation is as follows:

$$^{232}Th + n \rightarrow \ldots (^{233}Th, ^{233}Pa) \rightarrow ^{233}U \rightarrow \ldots \qquad (1.1)$$

The use of thorium for an industrial-scale nuclear fuel cycle is the subject of much current research, particularly in the USA and India (which has large Th deposits). Spent uranium-thorium fuel contains an insignificant amount of Pu, Np, Am and Cm compared to spent uranium fuel. So far only India is developing the thorium nuclear fuel cycle and operates an experimental thorium-uranium reactor.

Regular unirradiated or "fresh" uranium nuclear fuel used in different reactors such as Pressurised Water Reactor (PWR), Light Water Reactor (LWR), Channel-type Uranium Graphite Reactor (RBMK) and special nuclear submarine reactors typically consist of uranium dioxide, UO_2. An exception is the fuel of the UK's first-generation Magnox reactors, which used natural (i.e., non-enriched) metallic uranium encased in a cladding alloy, Magnox, consisting of Mg with additives of Al (0.8 wt.%) and Be (0.005 wt.%). The uranium fuel consists of two main isotopes: ^{238}U and ^{235}U; the ^{235}U is basically responsible for supporting the nuclear chain reaction.

One slow neutron (with energy less than 100 eV) might cause fission of ^{235}U nuclei into two smaller fission products (FPs) and release of two or three neutrons (on average 2.5), which can be captured and split other ^{235}U nuclei.

$$^{235}U + n \rightarrow {}^{236}U \rightarrow FPs + particles + energy \ (197 \ MeV). \quad (1.2)$$

Nuclear explosive with energy yield equal to 17 kilotonnes of trinitrotoluene (TNT) capability is provided by fission of 1 kg ^{235}U. However, the real energy yield from a nuclear explosion is less than this. The atomic bomb dropped on the Japanese city of Hiroshima contained 60 kilograms of highly-enriched uranium but the explosive energy was equal to 15 kilotonnes of TNT.

Neutrons with energy from 5 eV to several MeV are effectively absorbed by ^{238}U and cause transmutation of ^{238}U to plutonium isotopes. One of these reactions produces ^{239}Pu:

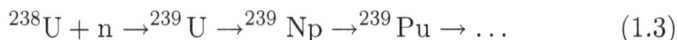

$$^{238}U + n \rightarrow {}^{239}U \rightarrow {}^{239}Np \rightarrow {}^{239}Pu \rightarrow \ldots \quad (1.3)$$

Neutrons with energy less than 5 eV are scattered by ^{238}U nuclei and move further. To initiate and support a permanent chain reaction in $(^{238}U, {}^{235}U)$-oxide fuel it is necessary to provide a suitable number of (preferably) slow neutrons as, for steady state, one of the neutrons released from the decay of ^{235}U nuclei should survive to split another ^{235}U nucleus. An effective nuclear reactor requires three features:

(1) Fuel with a suitable ^{235}U$/^{238}$U ratio or enrichment by ^{235}U;
(2) Critical volume or so-called "critical mass" of suitable radionuclide-containing fuel (IRSN Final Report, 2001), which should be sufficient to support the fission chain reaction, taking into account that some neutrons leave the reactor volume without splitting ^{235}U nuclei;
(3) The use of a neutron moderator (e.g., heavy water, graphite, water) to provide neutrons with energy less than 5 eV and to avoid intensive neutron absorption by ^{238}U.

Uranium-235 is the only fissile isotope that might be extracted in significant amounts from naturally occurring ores. The current ^{235}U content is approximately 1 kg in 140 kg (0.72 wt.%) of natural uranium, which is mainly ^{238}U with admixture of ^{235}U and ^{234}U. Two billion years ago, ^{235}U content in natural uranium was 3–4 wt.%. This caused conditions that led to a "natural nuclear reactor" in Oklo, Gabon, Africa (Gauthier-Lafaye *et al.*, 1996). Separation of ^{235}U from natural uranium is a complicated and costly process. Therefore, highly enriched fuel (with greater than 80% ^{235}U content) has limited applications in, for example, the compact reactors of nuclear submarines. Common enrichment of uranium oxide fuel varies from 2 to 8 wt.%. A typical composition of low-enriched uranium in nuclear fuel is as follows (in wt.%): ^{238}U-95.00; ^{235}U-4.05; ^{234}U-0.95 (King and Putte, 2003).

During reactor operation approximately 70–75 wt.% ^{235}U and only 2–3 wt.% ^{238}U burns to create new actinides such as Pu, Np, Am and Cm along with other artificial radionuclides. Most (80 wt.%) of the ^{235}U consumed is burned by fission reactions and the minor part (20 wt.%) by neutron capture to ^{236}U. Newly formed ^{236}U is then burned by neutron capture to ^{237}Np, which is then converted in part (25 wt.%) to ^{238}Pu. Most (93 wt.%) of the ^{238}U consumed is burned by neutron capture into ^{239}Pu and only 7% by fast neutron fission. Figure 1.3.1 shows typical nuclear reactions in a reactor as a function of time (Cohen, 1977).

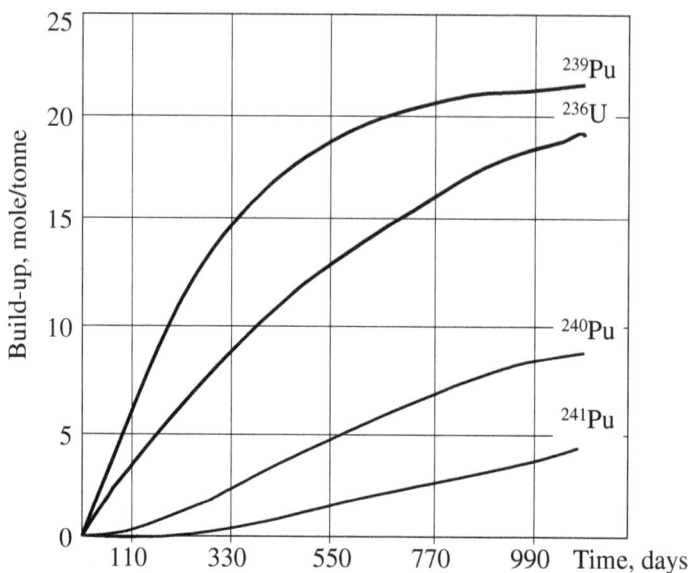

Fig. 1.3.1.   Build-up of the main actinides as a function of time (days) in a reactor core (after Cohen, 1977). To estimate actinide concentrations in ppm, data from this figure should be multiplied by the actinide atomic weight.

Actinide contents in spent nuclear fuel (SNF) depend on the level of uranium enrichment, the reactor type and the extent of fuel burn-up. One metric tonne of SNF contains (in kg): uranium oxide — 970–930; Pu — 5–10; Np — 0.2–0.8; Am — 0.2–2.1; Cm — 0.01–0.2. Annually, one 1,000 MW LWR produces approximately 1 metric tonne of Pu and 30 kg in total of so-called "minor" actinides (MA) such as Np, Am and Cm. Plutonium, therefore, is the main actinide formed as a result of burning uranium oxide fuel. A simplified schedule of uranium transmutation in nuclear reactors might be as follows:

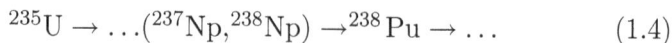

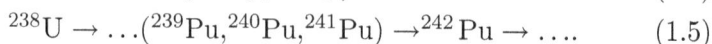

$$^{235}U \rightarrow \dots (^{237}Np, ^{238}Np) \rightarrow ^{238}Pu \rightarrow \dots \tag{1.4}$$

$$^{238}U \rightarrow \dots (^{239}Pu, ^{240}Pu, ^{241}Pu) \rightarrow ^{242}Pu \rightarrow \dots . \tag{1.5}$$

Neutron irradiation of ^{239}Pu causes its transmutation into new Pu isotopes and other actinides. Figure 1.3.2 shows the build-up of actinides on irradiation of ^{239}Pu as a function of time.

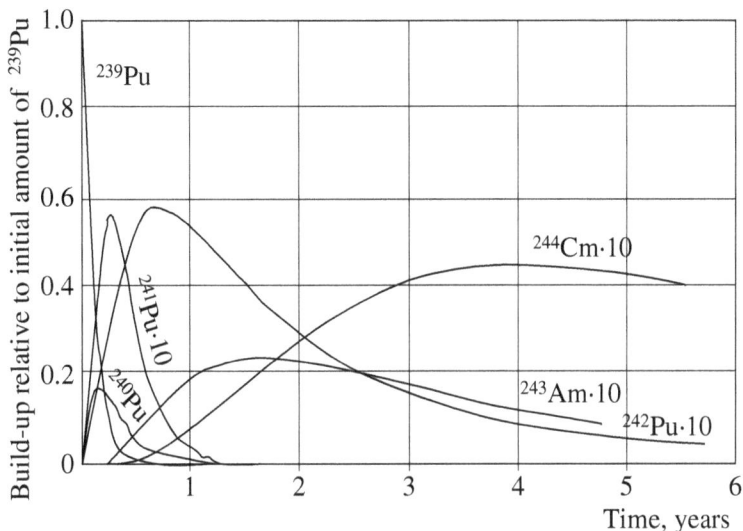

Fig. 1.3.2.   Build-up of actinides as a function of time on irradiation of ^{239}Pu by neutron flux $3 \ 10^{14}$ n/cm^2 sec (after Hyde and Seaborg, 1957).

On reprocessing, SNF is dissolved in nitric acid followed by extraction of uranium and plutonium for further reuse by the nuclear industry or temporary storage. Minor actinides (MA), such as Np, Am and Cm and other newly-formed radionuclides (fission products), form approximately 3–5 wt.% of SNF and are typically immobilised in a vitreous wasteform to give vitrified high level radioactive wastes (HLW).

1.3.2. *Weapons-grade plutonium*

Although the isotopic composition of plutonium extracted from SNF varies over a broad range depending on the type of reactor and the fuel burn-up, it can still be used in nuclear weapons. The total world plutonium inventory is not reported but a rough calculation indicates at least 2,000 metric tonnes at the beginning of the 21st century. One 1,000 MW thermal reactor (such as LWR, RBMK) produces approximately 1 metric tonne of Pu per year — the resulting annual increase in the world's plutonium inventory is at least 100 metric tonnes.

The preferable plutonium isotope used in modern warheads is
^{239}Pu. Complete fission of 55.6 g ^{239}Pu is accompanied by energy
release equivalent to 1,000 tonnes of TNT although the actual energy
release achieved by a nuclear explosion is much less. For example, the
plutonium bomb "Fat Man" dropped on Nagasaki in 1945 contained
6.1 kg Pu, but the explosive yield was 21 kilotonnes of TNT. One
control on the power of any nuclear weapon is the ^{240}Pu content
which generates neutrons through spontaneous fission. Such neutrons
may initiate a chain reaction too quickly to allow efficient detonation
of the nuclear warhead. Admixture of more than 7 wt.% ^{240}Pu in a
plutonium warhead makes the nuclear explosion unpredictable. The
isotopic composition of plutonium extracted from regular SNF is not
optimal for high-power nuclear weapons. Therefore, production of
weapon plutonium with less than 7 wt.% ^{240}Pu content is performed
only in special reactors, which are typically graphite-moderated using
gas or water cooling at atmospheric pressure and are capable of fuel
element exchange while on line — e.g., at a lower fuel burn-up when
the ratio of ^{240}Pu/^{239}Pu is relatively low (see Fig. 1.3.1).

^{238}Pu is 300 times more alpha-radioactive than ^{239}Pu
(Table 1.1.2). This isotope has been considered a completely "civilian' material. It is used as a heat source in, for example, radioisotope
thermoelectric generators (RTG), "nuclear" batteries of heart pacemakers and in accelerated radiation damage research. However, the
extremely high radiotoxicity of ^{238}Pu is a serious concern. Plutonium-
238 (as well as other plutonium isotopes) is a potential "dirty" bomb
material for possible terrorist attacks. ^{238}Pu is used in sealed radioactive sources utilised in applications on many sites operated by small
users. ^{238}Pu is also used as a heat source and in RTGs to produce
electricity for unmanned spacecraft and interplanetary probes. The
Russian classification of plutonium is given in Table 1.3.1.

It is important to note that modern nuclear weapon warheads consist not of pure metallic plutonium but its alloy, with approximately
1–5 wt.% Ga in order to stabilise the delta-polymorph (δ-Pu). Gallium admixture regularly accompanies Pu-containing materials and
wastes, and it makes direct conversion of weapon plutonium into
MOX or Pu-ceramic fuel for transmutation reactions difficult. One

Table 1.3.1.   Plutonium classification (Stukin and Bystrova, 2003).

Plutonium type	Content of critical isotope (in wt.%)
Super pure (main isotope ^{239}Pu)	^{240}Pu $\leq$ 2–3
Weapon (main isotope ^{239}Pu)	2–3 $<^{240}$Pu < 7
Fuel-type (main isotope ^{239}Pu)	$7 \leq^{240}$Pu ≤ 19
Reactor-type (main isotope ^{239}Pu)	^{240}Pu >19
Industrial	$20 \leq^{238}$Pu ≤ 80
Medical	^{238}Pu > 80

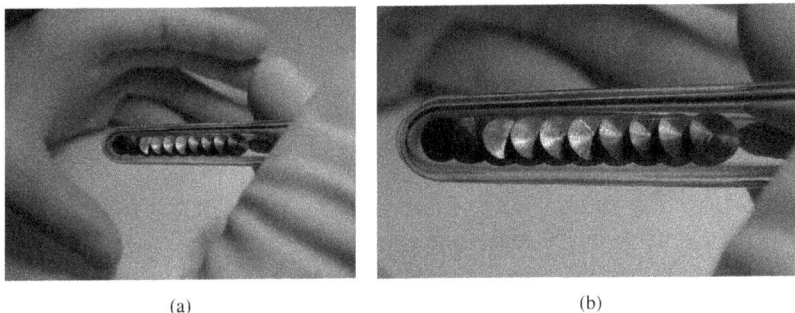

(a)   (b)

Fig. 1.3.3.   Pellets of pure metallic plutonium for laboratory use: each pellet is 5 mm in diameter and weighs approximately 0.5 g. Grey-blue oxide film covering the surface of plutonium metal can be clearly observed. Glass capsule is filled with inert gas and sealed for safe storage.

kilogram of typical weapon Pu accumulates almost 0.2 litres of helium; 3.7 grams of americium and 1.7 grams of uranium over a 50-year period. Metallic plutonium (Fig. 1.3.3) is chemically more active than metallic uranium. It slowly oxidises in dry air at 50°C acquiring a film of PuO_2.

Air with high relative humidity accelerates Pu oxidation leading to formation of a porous PuO_2 layer on the metal surface. In moist air plutonium corrodes 200 times faster than in dry air at room temperature and 100,000 times faster at 100°C. Oxidation of plutonium in humid argon is faster than in humid air, indicating it is caused mainly by interaction with water, not air. Monolithic pieces of metallic plutonium (several mm size) ignite spontaneously in air at 300–500°C (depending on source); however, fine plutonium powder may ignite at 150–200°C.

32   *Crystalline Materials for Actinide Immobilisation*

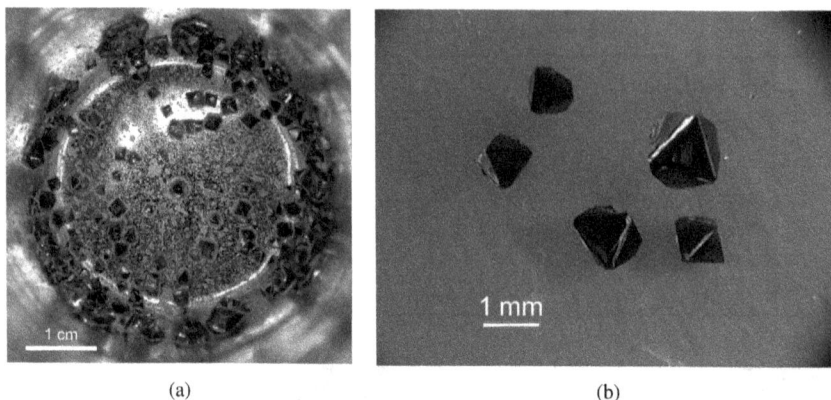

(a)   (b)

Fig. 1.3.4.   Single crystals of PuO_2 obtained at the V.G. Khlopin Radium Institute by flux method: at the bottom of platinum crucible immediately after synthesis (a) and extracted after rinsing in ammonium hydroxide and nitric acid (b) the PuO_2 synthesised unit cell parameter was a = 5.397(1).

The common plutonium isotopes decay primarily by emitting $\sim$5 MeV alpha-particles. However, one gramme of ^{238}Pu generates more heat (0.56 W) than the same amount of ^{239}Pu (0.002 W) and ^{242}Pu (0.0001 W). As a result, the surface temperature of a sealed stainless steel capsule containing $^{238}PuO_2$ is above room temperature.

Plutonium dioxide, PuO_2, is often found in plutonium wastes or from conversion of metallic plutonium from dismantled warheads. Stoichiometric PuO_2 is obtained after sintering in air at temperatures above 1,250C. Single crystals of PuO_2 1–2 mm in size (Fig. 1.3.4) to be used as standards (for electron microprobe analysis, EPMA and X-ray diffraction, XRD) can be obtained by a flux method (see Sec. 4.7).

Sintered PuO_2 is a relatively inert material that is difficult to dissolve in acids, behaviour which should be taken into account when developing starting precursor fabrication techniques for ceramic synthesis. Self-irradiation of sintered PuO_2 during long-term storage can decrease its chemical durability. Rapid dissolution of 20-year old pellets of sintered ^{238}Pu-oxide in nitric acid — behaviour never observed under the same conditions for pellets of sintered ^{239}Pu-oxide — has been observed (Zirlin et al., 2008). A hundred-fold increase in dissolution rate in distilled water has been reported (Patterson *et al.*, 1974) for $^{238}PuO_2$ as compared to the rate for $^{239}PuO_2$. The biological behaviour of various types of PuO_2 is also different: sintered $^{239}PuO_2$ (if inhaled) does not leave the lungs in contrast to $^{238}PuO_2$ and unsintered $^{239}PuO_2$, which are highly mobile and move from the lungs to the blood stream.

Dissolution of PuO_2 may be encouraged by steam pre-treatment in inert gas at 200–500°C or by adding hydrofluoric acid, HF, to boiling nitric acid. The highest concentration of plutonium achievable in aqueous nitrate solution is 150 g/l.

A serious difficulty in handling plutonium is related to the criticality problem. The critical mass of pure ^{239}Pu without it being surrounded by other materials is 10 kg and 40 kg for ^{240}Pu (Mark, 1993). However, these masses may be smaller in the presence of neutron reflectors. The critical amount of ^{239}Pu is 200 g in solution or approximately 1 kg ^{239}Pu in pure metal or alloy (Fishlock, 2005). Industrial treatment of plutonium is not done in more than 4.8-litre vessels, or of plutonium solutions with plutonium content more than 8 g per litre (Fishlock, 2005).

1.3.3. *Minor actinides*

1.3.3.1. *Neptunium-237*

Neptunium-237 is one of the most important long-lived minor actinides (MA) to be involved in immobilisation processes. It accumulates in different types of SNF (which typically contain hundreds of grammes of Np in one metric tonne of SNF) as a result of ^{235}U transmutation. Also, ^{237}Np is formed as a result of ^{241}Am decay. The half-life of ^{237}Np exceeds 2 million years (Table 1.1.2) making neptunium disposal difficult. Neptunium, like other actinides, is reactive and forms compounds with many other elements, such as halides, oxygen and hydrogen. Neptunium is far more mobile in the environment than plutonium — e.g., neptunium is soluble and tends to remain in the groundwater unlike other actinides such as plutonium and americium, which are absorbed by the soil.

34   *Crystalline Materials for Actinide Immobilisation*

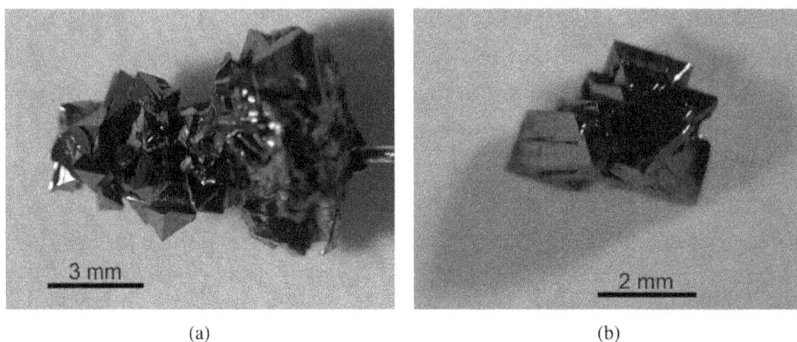

(a)   (b)

Fig. 1.3.5.   Single crystals of NpO_2 obtained at the V.G. Khlopin Radium Institute by flux method: aggregate of crystals grown from the seeds on a platinum
wire (a) and single crystal after rinsing in ammonium hydroxide and nitric acid
(b) The NpO_2 unit cell parameter was a = 5.433(1).

Neptunium can be used as a component in neutron detection
equipment. Typical forms of neptunium that need immobilisation are
aqueous nitrate solutions (from SNF reprocessing) and oxides. Low-
temperature calcination (at 275–450°C) of neptunium nitrates in air
forms Np_3O_8, which is similar to U_3O_8. Stoichiometric NpO_2 may be
synthesised by sintering neptunium salts (with any Np valence state)
in air at 700–800°C. Single crystals of NpO_2, 1–4 mm in size, can be
obtained by a flux method (Fig. 1.3.5). Sintered NpO_2 (similar to sintered PuO_2) is hard to dissolve in acids, although Np_3O_8 is easily dissolved by nitric acid. Hot H_2SO_4 is frequently used to dissolve NpO_2.

1.3.3.2. *Americium*

The two main americium isotopes, ^{241}Am and ^{243}Am, accumulate
as intergrowths (admixtures) in different types of SNF (from hundreds of grammes to several kilogrammes Am in one metric tonne of
SNF). The ^{241}Am/^{243}Am ratio depends on the reactor type and fuel
burn-up. In general, ^{241}Am content is approximately two times less
than ^{243}Am in SNF from thermal reactors but two times higher than
^{243}Am in SNF from fast reactors.

Americium-241 is a result of ^{238}U transmutation into plutonium
isotopes, including ^{241}Pu which decays to ^{241}Am. Macro-quantities

of ^{241}Am can be chemically separated from purified irradiated plutonium after several years' exposure. Americium-241 also accumulates
in aged metallic warhead plutonium, which always contain admixtures of ^{241}Pu.

^{241}Am in small quantities (typically less than 35 kBq) is used as
a source of ionising radiation in domestic smoke detectors. It is also
used in radiography — in gauges to measure the thickness of flat
glasses, for example.

Americium-243 arises from neutron irradiation of ^{241}Am or ^{242}Pu.
Its significantly longer half-life than ^{241}Am (7,380 years compared to
432 years) and its lower radiotoxicity make this isotope quite attractive for laboratory and other uses. This is why ^{243}Am is artificially
produced in macro-quantities by neutron irradiation of pure ^{242}Pu.

In aqueous solution, americium (similar to lanthanides) usually occurs in a stable 3+ valence state and has similar chemical
behaviour to tri-valent lanthanides. Sintering in air of americium
nitrate, oxalate and other salts forms AmO_2 with fluorite-type cubic
structure (similar to ThO_2, UO_2, NpO_2 and PuO_2). Sintering of
AmO_2 in reducing atmosphere (in the presence of hydrogen) leads to
two modifications of Am_2O_3: at 600°C — cubic (similar to Mn_2O_3),
and at 800°C — hexagonal (similar to La_2O_3).

1.3.3.3. *Curium*

Curium accumulates in SNF to the least extent of the MA (from tens
to hundreds of grammes Cm per metric tonne of SNF). Curium-
242 and 244 result from neutron irradiation of ^{241}Am and ^{239}Pu,
respectively:

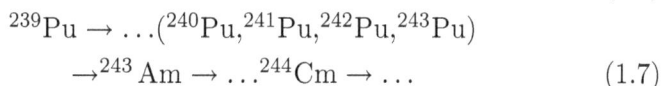

$$^{241}Am \rightarrow ...^{242}Cm \rightarrow ^{243}Cm \rightarrow ... \qquad (1.6)$$
$$^{239}Pu \rightarrow ...(^{240}Pu, ^{241}Pu, ^{242}Pu, ^{243}Pu)$$
$$\rightarrow ^{243}Am \rightarrow ...^{244}Cm \rightarrow ... \qquad (1.7)$$

Long-term disposal of most curium isotopes is not a concern
due to their short half-lives (see Table 1.1.2). However, two long-
lived curium isotopes — ^{248}Cm and ^{247}Cm (with half-lives $4.7 \cdot 10^5$
and $1.6 \cdot 10^7$ years, respectively — have prospective uses in RTGs

although how to obtain these isotopes in macro-quantities is unclear. One gramme of ^{244}Cm generates 3 W of power and one gramme of ^{242}Cm — 120 W sufficient for possible use in thermal electric generators.

In aqueous solutions, curium (similar to americium and lanthanides) occurs in the stable valence 3+ state.

Sintering curium oxalate in air at 650°C forms CmO_2 with cubic fluorite-type structure (similar to ThO_2, UO_2, NpO_2, PuO_2 and AmO_2). Sintering of CmO_2 in reducing atmosphere (in the presence of hydrogen) causes formation of cubic Cm_2O_3 which, after several weeks, transforms into a hexagonal polymorph (similar to La_2O_3).

1.3.3.4. *Berkelium and Californium*

Isotopes of Bk and Cf result from long in-reactor irradiation of ^{239}Pu. Most of their isotopes are short-lived and their levels in SNF of insignificant concern for immobilisation. Applications of californium include neutron start-up sources, calibrating instrumentation, treatment of certain cervical and brain cancers where other radiation therapy is ineffective, radiography of aircraft to detect metal fatigue, airport neutron-activation detectors of explosives, portable metal detectors, neutron moisture gauges in the oil industry and portable neutron sources. ^{251}Cf (half-life 898 years) might find some special uses, such as in explosive weaponry. ^{252}Cf and ^{254}Cf are strong sources of secondary neutrons and have been suggested for treatment of cancer tumours (Vdovenko, 1969).

Berkelium oxide BkO_2 has a cubic fluorite-type structure (similar to ThO_2, UO_2, NpO_2, PuO_2 and AmO_2). Californium oxide Cf_2O_3 has a monoclinic structure similar to Sm_2O_3.

1.4. Actinide Host-Phases

1.4.1. *Natural accessory minerals*

Accessory minerals are minor minerals, which occur in natural rocks in negligible amounts (less than 1 wt.%). Not all accessory minerals are chemically and mechanically durable but some demonstrate extremely high stability over long periods of time. Isotope geochronology provides dating of the oldest rocks and meteorites using accessory minerals such as zircon and monazite as the unique "chronometers" which preserve internal information about rock formation for billions of years. Some chemically-stable accessory minerals are intrinsic phases of natural radionuclides — e.g., thorianite, $(Th,U)O_2$ and thorite, $(Th,U)SiO_4$. Others accommodate natural actinides in their crystalline structures in the form of solid solutions such as zircon, $(Zr,U,Th)SiO_4$ and monazite, $(Ce,REE,Th,U)PO_4$, where REE are rare earth (mainly Ce and La — in natural monazite). The main commercial source of Th is monazite. Samples of various actinide-bearing natural accessory minerals of different age and accumulated dose of self-irradiation are available from mineralogical collections around the world.

Some geological processes may cause accumulation of accessory minerals. For example, formation of pegmatite veins in granite and other rocks is accompanied by an increase in accessory mineral concentration compared to the average content in a host massif. Pegmatites are well-known sources of large crystals of zircon, tourmaline and uraninite (Fig. 1.2.1), thorianite, beryl, apatite, pyrochlore, monazite, xenotime, garnet and other accessory minerals. In some cases, pegmatites are mined industrially to access the accessory minerals. Another important process affecting accumulation of accessory minerals is rock weathering. In general, most accessory minerals are more stable than rock-forming minerals (e.g., quartz, feldspars and mica). As a result of weathering the accessory minerals released from host rock are moved by water streams and may accumulate in so-called "placer deposits" (Fig. 1.4.1). Approximately 90% of the world zirconium production comes from zircon concentrate placers.

It is important to study accessory minerals because they may give pointers to potential host-phases for the immobilisation of actinides (and other long-lived radionuclides) into ceramic wasteforms and further disposal in particular geological formations. Such investigations should cover:

(1) Selection of durable (chemically and mechanically) host-phases with ability to incorporate actinides (and possibly other long-lived radionuclides) in their crystalline structures;

38   *Crystalline Materials for Actinide Immobilisation*

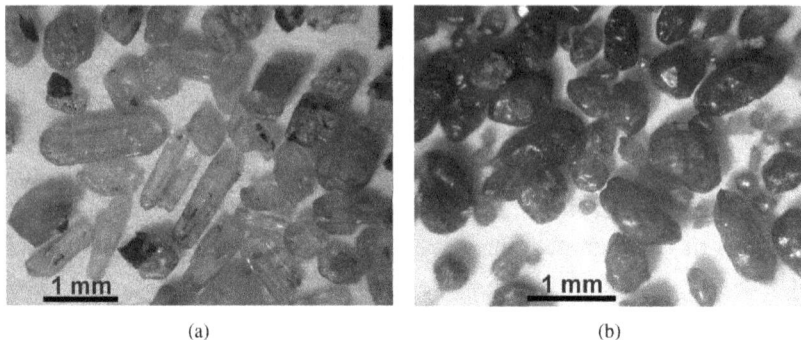

(a)   (b)

Fig. 1.4.1.   Crystals of natural monazite (a) and zircon (b) from placers (Ukraine).

(2) Geochemical compatibility of host-phases with proposed geological formation for the repository. If ceramic wasteforms are analogues of accessory minerals typical for a particular rock they should help preserve geochemical equilibrium in this formation (geological massif) after disposal of actinide wastes (Fig. 1.4.2);

(3) Resistance of selected host-phase to self-irradiation damage and chemical alteration processes over long periods of time.

In the remainder of this chapter we consider some minerals potentially capable of hosting actinides in their structures.

1.4.2. *Zircon and hafnon*

Zircon and hafnon are silicate minerals with zircon-type tetragonal structure. They represent complete solid solution between $ZrSiO_4$ and $HfSiO_4$. The name "zircon" originates from the Arabic and Persian languages and means "gold colour". "Hafnon" is created from the element hafnium as zircon derives from zirconium. Zircon is a well-known accessory mineral which occurs in different rock types (Fig. 1.4.3), including Lunar rock (Smith, 1974), placers (Fig. 1.4.1), pegmatites and meteorites. It is the main industrial source of zirconium and hafnium. At the same time, zircon is a popular gemstone (Fig. 1.4.3).

The mineral hafnon is defined as having a HfO_2 content in solid solution $(Hf,Zr)SiO_4$ exceeding 20 wt.% and is extremely rare. Natural hafnon with HfO_2 content 70–73 wt.% has been found only

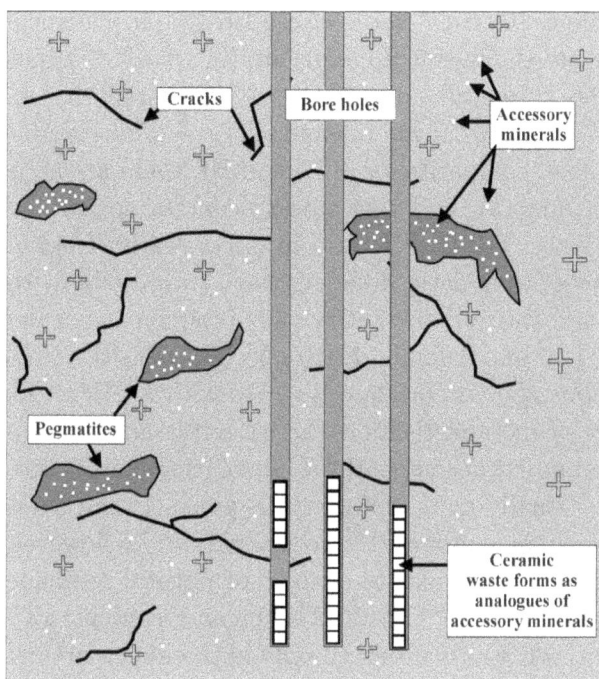

Fig. 1.4.2. Disposal of actinide wastes into boreholes in granite massif. The ceramic wasteforms are analogues of naturally-occurring accessory minerals typical in this type of rock. The repository site is considered an analogue of pegmatite veins with accumulated accessory minerals (Burakov and Anderson, 2001).

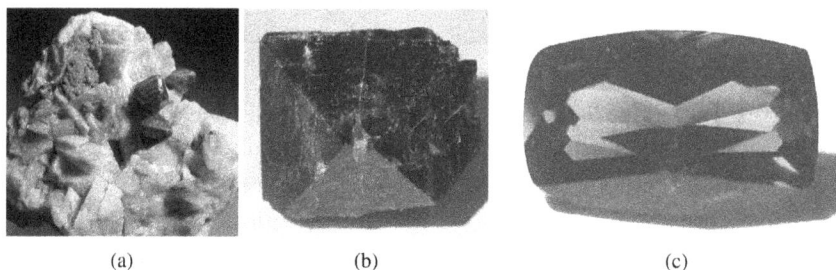

(a)   (b)   (c)

Fig. 1.4.3. Natural zircon: (a) crystals 8–10 mm in size in carbonatite rock (Kovdor deposit, Kola Peninsula, Russia); (b) single crystal 15 × 19 mm from albitite rock (Kola Peninsula, Russia); (c) transparent facet-cut 5 × 8 mm zircon gemstone (jacinth).

in Mozambique (Correia Neves *et al.*, 1974). The chemical compositions of the stoichiometric minerals are (in wt.%) — zircon: ZrO_2 —
67.2 (or 49.8 Zr metal) and SiO_2 — 32.8 (or 15.3 Si); and hafnon:
HfO_2 — 77.8 (or 66 Hf metal) and SiO_2 — 22.2 (or 10.4 Si). Due
to the chemical compatibility of Zr and Hf, there are no pure zircon
and hafnon minerals and they always contain admixtures of Hf or
Zr, respectively. The Zr/Hf ratio in zircon varies from 71 to 36, a
consequence of magmatic differentiation (from gabbro to granites)
(Kosterin and Zuyev, 1958). Crystalline zircon has a density from
~4.6–4.7 g/cm^3 and hafnon ~7.0 g/cm^3. The densities of $(Zr,Hf)SiO_4$
solid solutions should correlate directly with Zr/Hf ratio, however:
the process of metamictisation (under self-irradiation from admixtures of Th and U) decreases the solid solution density (to 3.9 g/cm^3
for zircon). Usually metamict zircon is green. U and Th contents in
natural zircon vary from 10^{-4} to several wt.%; however, for crystalline zircons the admixture content of natural radionuclides does
not usually exceed 10^{-2} wt.%. The highest contents of Th and U
(more than 1 wt.%) are observed only in metamict zircons and often
enrichment of zircon with these radionuclides is accompanied by low
Zr content ($\leq$50 wt.% ZrO_2) and high content of H_2O (to 10 wt.%),
P_2O_5 (to several wt.%), Fe_2O_3 (to 3–4 wt.%), CaO (to 4 wt.%) and
some other admixtures (Zubkov, 1989). It is assumed elements such
as the rare earths, Y, Ca, Fe and Na, may substitute for Zr but P
can partly occupy Si sites in the zircon structure. The absence of
natural crystalline zircons with high contents of these admixtures,
however, limits precise investigation of the capacity of the zircon lat-
tice to incorporate these elements. Natural crystalline zircon is chemically durable. The process of metamictisation under self-irradiation
from admixtures of Th and U decreases its chemical resistance. However, even completely metamict zircon demonstrates some durability
in acids and alkalis. Accessory zircon is a unique and well-known
material for use in isotope geochronology. The oldest natural rocks
so far measured (at 4 billion years) were characterised by isotope
analysis of zircon samples (Maas *et al.*, 1992). Zircon's mechanical
durability is high and its hardness is 7.5 on Mohs' scale. Zircon's

thermal stability exceeds $1,600°C$, enabling its industrial application in refractory ceramics (Simuya, 1989).

1.4.3. *Monazite*

Natural monazite is a well-known monoclinic phosphate of Ce and other rare-earths, $(Ce,RE,Th,U,Ca)PO_4$. Its name originates from Greek and means "to be solitary", referring to its somewhat rare occurrence (Mitchell, 1979). Monazite is an important ore for Th, Ce and other rare earths. It is a typical accessory mineral in different igneous and metamorphic rocks, pegmatites and placers (see Fig. 1.4.1). The chemical composition of monazites covers a broad range. They may contain (in wt.%): Ce_2O_3 — 26–31; P_2O_5 — 22–31; total oxides of La, Nd, Pr — 21–30; Y_2O_3 — 0.5–3.5; ThO_2 — 1–28; ZrO_2 — up to 7; SiO_2 — up to 6. In contrast with thorium admixtures of uranium and thorium in monazite are unusual. However, monazites containing approximately 16 wt.% UO_2 but only 11 wt.% ThO_2 (Gramaccioli and Segalstad, 1978), and 7 wt.% UO_2 but only 4 wt.% ThO_2 (Hutton, 1951) have been reported. Monazite's hardness is similar to silicate glass (5.0–5.5 on Mohs' scale) and its density is 4.6–5.4 g/cm^3. The chemical durability of monazite is lower than that of zircon; it dissolves slowly in hot HCl. At the same time, monazites even with high Th and U contents remain crystalline, whereas zircon species are metamict at much lower natural radionuclide contents. Accessory monazite, like zircon, is widely used for isotope geochronology.

1.4.4. *Zirconolite*

Zirconolite or zirkelite is a relatively rare mineral, usually with monoclinic structure and simplified formula $CaZrTi_2O_7$. The name "zirkelite" was dedicated to the memory of German petrologist F. Zirkel (1838–1912) (Mitchell, 1979). "Zirconolite", however, is confusing and probably a misnomer (Fleischer, 1975). Surprisingly, "zirconolite" has been accepted by most of the researchers developing ceramic nuclear wasteforms. Natural zirconolite may incorporate (in wt.%): UO_2 — 24; ThO_2 — 22 and RE_2O_3 — 32 (Lumpkin *et al.*, 2004).

Zirkelite (zirconolite) found in Lunar rock contains (on average, wt.%): ZrO_2 — 40; TiO_2 — 35; CaO — 9; UO_2 — 0.2; ThO_2 — 0.5; Y_2O_3 — 4; FeO — 6 (Busche *et al.*, 1972). Zirconolite with high Th and U content is usually metamict, however, some crystalline species with up to 11 wt.% UO_2 and 10 wt.% ThO_2 have been discovered (Bellatreccia *et al.*, 2002). Metamict species become crystalline (with monoclinic structure) after sintering at 1,100–1,200°C. Substitution of RE, U, Th for Ca and Nb, Fe, Mg for Ti causes occurrence of different polytypic forms: monoclinic 2M or 4M; orthorhombic and hexagonal (Lumpkin *et al.*, 2004). Zirconolite's hardness is 6 on Mohs' scale and its density varies from 4.5 to 5.1 g/cm^3. There is little information about its geochemical durability.

1.4.5. *Baddeleyite (monoclinic zirconia)*

Baddeleyite is natural monoclinic zirconia, ZrO_2. This mineral was named in memory of J. Baddeley, who discovered it in Sri Lanka (Mitchell, 1979). Together with zircon, baddeleyite is the major source of zirconium for industrial use (Fig. 1.4.4). It accompanies zircon in different types of rocks, including Lunar rocks (Smith, 1974), and placers. However, this mineral is much rarer than zircon. Little information is available about the radionuclide content in baddeleyite

Fig. 1.4.4.   Industrial concentrates of baddeleyite from the Kovdor deposit (Kola Peninsula, Russia).

samples and its behaviour under self-irradiation by natural radionuclides over long times. Baddeleyite has been found (Degueldre and
Hellwig, 2003) in uranpyrochlore phase, demonstrating a high resistance to chemical alteration and radiation damage. Some evidence of
amorphous domain formation was observed only at interfaces with
uranpyrochlore at estimated doses $(3–4.5) \cdot 10^{16}$ alpha-decay/mg. No
metamict or partly metamict baddeleyites are known.

An industrial concentrate of baddeleyite crystals from the Kovdor deposit (Fig. 1.4.4), studied at the V.G. Khlopin Institute by
Andrey Gedeonov, demonstrated bulk U content at levels 0.01 wt.%
and Th — 0.002 wt.%. Its hardness is 6.5 on Mohs' scale and the
density is 5.7 g/cm^3. At 1,250°C monoclinic ZrO_2 transforms into
tetragonal and then at 1,900°C it becomes hexagonal, and at 2,300°C
and higher it is cubic. The melting point of zirconia is $\sim$2,700°C so
its thermal stability is high but the behaviour of chemical admixtures incorporated in its crystalline structure on heating and their
influence on its polymorphism is unclear.

1.4.6. *Tazheranite (cubic zirconia)*

Natural cubic zirconia with simplified formula $(Zr,Ca,Ti,\ldots)O_2$ is
a rare mineral, discovered in 1966 in calciphyre (carbonate metasomatic rock) of the Tazheranskiy massif (Lake Baikal region, Russia). The name "tazheranite" was taken from the name of the massif
(Konev *et al.*, 1969). This zirconium mineral (72–73 wt.% ZrO_2)
contains admixtures of Ca and Ti (approximately 11 wt.% CaO
and 15–16 wt.% TiO_2), assumed responsible for stabilising the cubic
structure. The unit cell parameter is a = 5.100(3) Å and its density
5.01(2) g/cm^3.

There is no information available concerning the geochemical
durability of tazheranite and its natural radionuclide content.

1.4.7. *Xenotime*

Xenotime is a rare Y-phosphate with zircon-type tetragonal structure
and simplified formula YPO_4. Y_2O_3 and P_2O_5 contents in natural
samples vary from 55 to 63 and from 25 to 27 wt.%, respectively.

The name "xenotime" originates from the Greek and combines the words "foreigner" and "to revere" (Mitchell, 1979). For a long time, rare finds of xenotime crystals were not recognised. It is assumed there is very limited YPO_4-$ZrSiO_4$ solid solution. This may explain the partial substitution of Y for Th, U, Zr and P — for Si in natural samples. Xenotime hardness is relatively low (4–5) on Mohs' scale and its density is 4.3–4.7 g/cm^3. Radiation damage effects from Th and U in natural species of xenotime have yet to be studied in detail.

1.4.8. *Apatite*

The family of natural apatite includes a large number of hexagonal-structured minerals with general formula $A_5(BO_4)_3(F,Cl,OH)$, where "A" = Ba, Ca, Ce, K, Na, Pb, Sr and "B" = As, C, P, S, Si, V. "Apatite" is derived from Greek — "to deceive" (Mitchell, 1979). Identification of this mineral is difficult. It is often mistaken for other minerals. Widespread natural apatites are calcium-phosphate minerals (with CaO content — 54–56 wt.% and P_2O_5 — 41–42 wt.%) such as F-apatite, $Ca_5(PO_4)_3F$ (content of F up to 3.8 wt.%); Cl-apatite, $Ca_5(PO_4)_3Cl$ (content of Cl up to 6.8 wt.%) and hydroxyl-apatite, $Ca_5(PO_4)_3(OH)$. Usually, natural species contain variable amounts of F, Cl and hydroxyl grouped together. Apatite may also contain essential admixtures of (in wt.% — maximum content): MnO — 7; SrO — 24; RE_2O_3 — 12; Th — 0.6; SiO_2 — 2; SO_3 — 4; CO_2 – 5.

F-apatites are chemically the most durable and they are typical accessory minerals for different types of rock (Fig. 1.4.5), including Lunar rocks (Smith, 1974) and placers. Hydroxy-apatite is a well-known "biomineral", which makes up bones and teeth. The increased chemical durability of F-apatites explains the use of fluoride toothpastes and fluoridation of water supplies to prevent tooth decay: F replaces OH, increasing the tooth apatite resistance to dental caries. The hardness of Ca-phosphate F-Cl-apatite is relatively low and similar to silicate glass — 5 on Mohs' scale; its density 3.2–3.8 g/cm^3. All apatite species dissolve in HCl, H_2SO_4 and HNO_3. F-apatites and hydroxy-apatites are used extensively for fertiliser production. Radiation damage effects from Th and U in natural species of phosphate-apatite have not been studied in detail.

(a)   (b)

Fig. 1.4.5.   Samples of natural F-apatite: (a) apatite-nepheline rock — main light-green part of the stone consists of apatite grains (Hibin Mountains, Kola Peninsula, Russia); (b) gem-quality transparent crystal from calciphyre rock (Slyudyanka deposit, Lake Baikal region, Russia).

1.4.9. *Pyrochlore*

Pyrochlore is the common name for a large family of natural minerals with cubic pyrochlore-type structure. The name "pyrochlore" originates from the Greek words for "fire" and "green" because some pyrochlore samples become green after firing (Mitchell, 1979). The simplified formula of natural pyrochlore might be described as $A_2B_2O_6(OH,F)$, where "A" = Na, K, Ca, Sr, Ba, Fe^{2+}; RE, U, Th,...; "B" = Nb, Ta, Ti, Fe^{3+},... Pyrochlore contains 56–68 wt.% Nb_2O_3, but there are also Ta-rich species (Ta-pyrochlore or microlite) with 68–77 wt.% Ta_2O_5 and Ti-rich pyrochlore (betafite and Ti-pyrochlore) with 32–35 wt.% TiO_2 and (Nb_2O_3 + Ta_2O_5) less than 25 wt.%. Unit cell parameters and densities vary over a broad range depending on chemical composition, level of metamictisation and hydration. In general, a = 10.33–10.40 Å and density is from 3.8 to 6.4 g/cm^3. Often analyses of natural pyrochlores do not allow calculation of the stoichiometric formula. A site cation deficiency is typical (Van Wambeke, 1970) and it may exceed 20%. In some cases cation deficiency is correlated by IR-spectroscopy with H_3O^+

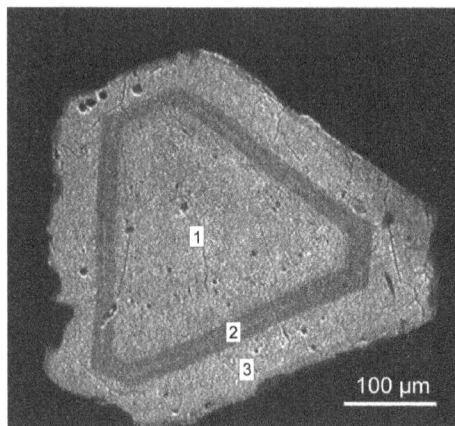

Fig. 1.4.6.   Backscattered electron SEM image of natural pyrochlore crystal, $(Na,Ca,Sr,Fe,U)2(Nb,Ta,Ti)_2O_6(OH)$ (Karelia, Russia). Crystal was mounted into acrylic resin. Uranium content (areas 1 and 3) is approximately 12 wt.% el., and in zone 2 — 5 wt.%el.

content. Pyrochlore structured compounds are characterised by ion-exchange from the A site (Belinskaya, 1984), leading to cation release under natural weathering (Van Wambeke, 1970).

Uranium contents in samples of natural pyrochlore range from 0 to 20–22 wt.% el, while uranium admixtures may vary over a broad range in the matrix of the same crystal (Fig. 1.4.6). Essential admixture of Th is not typical and only rare natural samples contain more than 1 wt.% Th.

Pyrochlore samples containing U and Th are often metamict and look like black (or dark-brown) glass (Fig. 1.4.7).

While some samples have been found with other accessory minerals in placers, pyrochlore is not observed in ancient placers or in placers located far away from the original pyrochlore deposits. This indicates natural pyrochlore has a relatively low geochemical durability.

1.4.10. *Perovskite*

Natural perovskite is a titanate mineral with rhombic (pseudo-cubic) crystal structure and simplified formula $CaTiO_3$ (41 wt.% CaO and

Fig. 1.4.7. Metamict samples of natural pyrochlore from granite pegmatite (Karelia, Russia).

58.8 wt.% TiO_2). "Perovskite" is named after the Russian mineralogist Count L.A. Perovskiy (1792–1856). Natural samples of perovskite contain different chemical admixtures such as Na, rare earths, Th, Sr, Mg, Fe^{2+} substituting in part for Ca; and Nb, Ta, Zr, Si and Sn^{4+} occupying some of the Ti positions. Admixture contents in perovskites such as loparite (Fig. 1.4.8) can achieve (in wt.%): rare earths (calculated in bulk oxide RE_2O_3) — 28–45; Nb_2O_5 — 26; SrO — 3; Fe_2O_3 — 3 and ThO_2 — 3. The hardness of perovskite is not high and similar to glass (5.5) on Mohs' scale and its density is 4.0–4.3 g/cm^3. Metamict species can contain up to 3.5 wt.% H_2O.

1.4.11. *Garnet*

The natural garnet family includes silicates with cubic structure and general formula $A_3B_2(SiO_4)_3$, where "A" = Ca, Fe^{2+}, Mg, Mn^{2+}; and "B" = Al, Cr, Fe^{3+}, Mn^{3+}, Ti, V, Zr. "Garnet" derives from the Latin and reflects the similarity between crystals of red-coloured garnet species (almandine, pyrope) and grains of pomegranate (*"granatum"* in Latin). Garnet species of different chemical composition have separate names such as: almandine, $Fe_3Al_2(SiO_4)_3$ (see Fig. 1.4.9); pyrope, $Mg_3Al_2(SiO_4)_3$; spessartine, $Mn_3Al_2(SiO_4)_3$; grossular, $Ca_3Al_2(SiO_4)_3$ (see Fig. 1.4.9); andradite, $Ca_3Fe_2(SiO_4)_3$; uvarovite,

Fig. 1.4.8. Twinned crystals (6–9 mm in size) of perovskite-loparite in syenite rock (Kola Peninsula, Russia).

(a)   (b)   (c)

Fig. 1.4.9. Crystals of natural garnet. (a) almandine 25×20 mm (Karelia, Russia); (b) grossular 25×19 mm (Viluy-river region, Eastern Siberia, Russia); and (c) uvarovite (1–2 mm in size) on chromite rock (Perm region, Russia).

$Ca_3Cr_2(SiO_4)_3$ (see Fig. 1.4.9); schorlomite, $Ca_3(Fe,Ti)_2(Si,Ti)_3O_{12}$; knorringite, $Mg_3Cr_2(SiO_4)_3$; kimzeyite, $Ca_3(Zr,Ti)_2(Si,Al)_3O_{12}$; goldmanite, $Ca_3(V,Al,Fe)_2(SiO_4)_3$. Pure garnets are rare and usually garnet is represented by limited or broad solid solutions in the systems: "almandine-pyrope"; "almandine-spessartine"; "grossular-andradite"; "grossular-uvarovite-schorlomite"; "pyrope-knorringite", etc. (Berry *et al.*, 1983). Density of garnet correlates directly with its chemical composition and varies from 3.5–3.6 (pyrope, grossular) to 4.2–4.3 g/cm^3 (spessartine, almandine).

Most garnet species are characterised by high mechanical durability: hardness of almandine, grossular and pyrope is 7.5 on Mohs' scale, but for andradite it is only 6.5–7.0. Transparent garnet varieties — in particular, pyrope and almandine — are famous gemstones. Placer deposits of pyrope and almandine are common, indirectly confirming their geochemical durability. A unique Th-U-bearing grossular containing 0.1–0.3 wt.% UO_2, up to 3.8 wt.% ThO_2 and 6–8 wt.% rare earth oxides has been reported (Kumral *et al.*, 2007). However, no information is yet available from studies of natural samples concerning the resistance of the garnet structure to radiation damage from natural radionuclides.

1.4.12. *Murataite*

Murataite is a rare mineral found in quartz-microcline pegmatites in Colorado, USA (Adams *et al.*, 1974). "Murataite" was named after the geochemist K.J. Murata. This cubic mineral (a = 14.86(1) Å) has density of around 4.64 g/cm^3 and complex chemical composition, which may be described by $(Na,Y,Er)_4(Zn,Fe)_3(Ti,Nb)_6O_{18}(F,OH)_4$. Approximate contents of the major elements (in wt.%): Na_2O — 6; TiO_2 — 38; ZnO — 12; Y_2O_3 — 12; RE_2O_3 — 10; Nb_2O_3 — 10; FeO — 4; F — 7. Murataite that does not contain Nb and F (and similar unit cell parameter a = 14.87(2) Å) was discovered in alkaline pegmatites of the Lake Baikal region of Russia (Portnov *et al.*, 1981). Its chemical composition was (in wt.%): Na_2O — 6; CaO — 7; TiO_2 — 48; ZnO — 16; Y_2O_3 —21; RE_2O_3 — 1; FeO — 2.

There is no information available concerning geochemical durability of murataite and any radionuclides in its structure.

1.4.13. *Kosnarite*

Kosnarite is a rare phosphate mineral with hexagonal structure and formula $KZr_2(PO_4)_3$, discovered in pegmatites (Brownfield *et al.*, 1993). The mineral is named after R.A. Kosnar of Black Hawk, Colorado, USA. Unit cell parameters are: a = 8.687(2); c = 23.877(7) Å. Density is approximately 3.2 g/cm^3. Contents of main elements (in wt.%): P_2O_5 — 43–44; ZrO_2 — 44–45; K_2O — 9; N_2O — 1–2.

There is no information available concerning the geochemical durability of kosnarite or its natural radionuclide content.

1.4.14. *Natural gels*

Natural U-bearing solid gels are unusual but have been found in uranium deposits in Kazakhstan (Smetannikov, 1997) and the Lake Baikal region of Russia (Aleshin *et al.*, 2007) and the Southern Urals area (Dimkov *et al.*, 2003). These gels contain U, Si plus water. In some cases, the gels contain mainly S and P instead of Si (Dimkov *et al.*, 2003). Gels formed as a result of metasomatic processes initially precipitated as amorphous materials. The survival of the gels in an open geochemical system for millions of years demonstrates their high chemical durability. Chemical composition of the gel from Kazakhstan (Helean *et al.*, 1997 and Burakov *et al.*, 2006) is relatively homogeneous except local zones surrounded inclusions of galena (Table 1.4.1, Fig. 1.4.10). No separate oxide phases of U, Zr or Si are observed in the gel matrix.

After sintering in air for 1 hour at $1,400°C$, the gel largely transformed into U-doped crystalline zircon, $(Zr,U)SiO_4$. After sintering, the uranium was not found in any other phases besides zircon.

The geochemical durability of such gels is the subject of further research. It seems likely the presence of underground water in the open geochemical system helped preserve the initial gel structure. In addition, it has been speculated (Burakov *et al.*, 2006) that the

Table 1.4.1. Chemical and radionuclide composition of natural Zr-U-Si-gel from Kazakhstan (Burakov *et al.*, 2006).

Elements in wt.% (except O and H) from EPMA						
Ca	Mg	Al	Zr	U	Fe	Si
1.5–2.0	0.5–1.2	0.5–2.5	20.9–26.5	3.6–12.8	0.7–13.8	8.0–10.4
Nuclides in Bq/g from gamma-spectroscopy						
^{238}U		^{235}U		^{226}Ra	^{228}Th	
1590 ± 199		99 ± 20		1590 ± 100	596 ± 99	

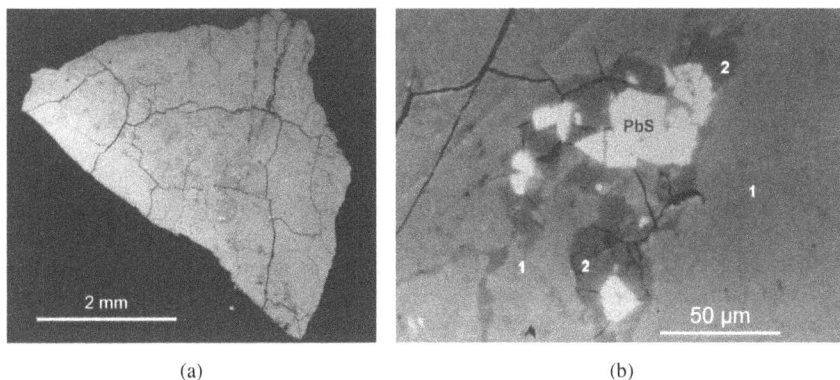

(a)   (b)

Fig. 1.4.10.   Backscattered electron SEM images of natural Zr-U-silicate gel: (a) general view and (b) detail of matrix (1) and local zones (2) with lower uranium content and inclusions of galena, PbS (Burakov *et al.*, 2006).

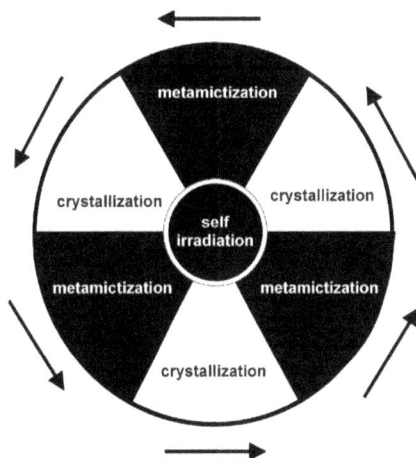

Fig. 1.4.11.   Schematic of Zr-silicate gel stability over long periods of time in the open geochemical system in the presence of underground water. Self-irradiation provides equilibrium between two competing processes: gel crystallisation largely into zircon and zircon metamictisation (Burakov *et al.*, 2006).

natural gels' high chemical durability is caused by two competing processes that exist under self-irradiation conditions: (1) crystallisation of the gel and (2) metamictisation of the crystallised zircon and other phases (Fig. 1.4.11).

52   *Crystalline Materials for Actinide Immobilisation*

References

Adams J.W., Botinelly T., Sharp W.N. and Robinson K. (1974) Murataite, a new complex oxide from El Paso County, Colorado, *Amer. Mineral.* **59**, 172–176.

Aleshin A.P., Velichkin V.I. and Krylova T.L. (2007) Genesis and formation conditions of deposits in the unique Strel'tsovka molybdenum-uranium ore field: New mineralogical, geochemical, and physicochemical evidence, *Geology of Ore Deposits* **49**, 392–412.

Belinskaya F.A. (1984) Ion exchange and isomorphism in pyrochlore-type compounds. In *Ion Exchange and Ionometry*, Vol. 4, Leningrad State University, pp. 3–13 (in Russian).

Bellatreccia F., Della Ventura G., Williams C.T., Lumpkin G.R., Smith K.L. and Colella M. (2002) Non-metamict zirconolite polytypes from the feldspathoids-bearing alkali-syenitic ejecta of the Vico volcanic complex (Latium, Italy), *Eur. J. Mineral.* **14**, 809–820.

Berdjis C.C. (ed). (1971) *Pathology of Irradiation* (Williams and Wilkins, Baltimore).

Berry L.G., Mason B. and Dietrich R.V. (1983) *Mineralogy. Concepts, Descriptions, Determinations*, 2nd edn., (W. H. Freeman and Company, San Francisco).

Brownfield M.E., Foord E.E., Sutley S.J. and Botinelly T. (1993) Kosnarite, $KZr_2(PO_4)_3$, a new mineral from Mount Mica and Black Mountain, Oxford County, Maine, *Amer. Mineral.* **78**, 653–656.

Burakov B.E and Anderson E.B. (2001) Crystalline ceramics developed for the immobilization of actinide wastes in Russia. *Proc. 8th Int. Conf. Radioactive Waste Management and Environmental Remediation ICEM'01*, Bruges, Belgium, 30 September–4 October 2001, CD, Session 39.

Burakov B.E., Smetannikov A.Ph. and Anderson E.B. (2006) Investigation of natural and artificial Zr-silicate gels. In *Scientific Basis for Nuclear Waste Management XXIX*, ed. P. Van Iseghem, *Mat. Res. Soc. Symp. Proc.* **932**, 1017–1024.

Busche F.D., Prinz M., Keil K. and Kurat G. (1972) Lunar zirkelite: A uranium-bearing phase, *Earth and Planetary Science Letters* **14**, 313–321.

Cherdintsev V.V., Kazachevskiy I.V., Sulerzhitskiy L.D. and Kuzmina E.A. (1965) About plutonium-239 in nature, *Geohimiya* **10**, 1257–1258 (in Russian).

Cohen B.L. (1977) High-level radioactive waste from light-water reactors *Rev. Mod. Phys.* **49**, 1–20.

Correia Neves J.M., Lopes N. and Sahama G. (1974) High hafnium members of the zircon-hafnon series from the granite pegmatites of Zambezia, Mozambique, *Contrib. Mineral. Petrolog.* **48**, 73–80.

Cotton F.A. and Wilkinson G. (1988) *Advanced Inorganic Chemistry*, 5th edn. (Wiley-Interscience, New York).

Degueldre C. and Hellwig Ch. (2003) Study of a zirconia based inert matrix fuel under irradiation, *J. Nucl. Mater.* **320**, 96–105.

Dimkov Yu.M., Doynikova O.A. and Volkov N.I. (2003) Find of U-Fe-Zr-Ti-S-P-gel in exogenous-epigenetic uranium deposit "Hohlovskoe" (Southern Trans-Urals Region), *Geochimiya* **11**, 1–9 (in Russian).

Fishlock D. (2005) Drama of plutonium, *Nucl. Eng. Int.* **50**, 42–43.

Fleischer M. (1975) New mineral names, *American Mineralogist* **60**, 340–341.

Flerov G.N. and Petrzhak K.A. (1940) Spontaneous fission of uranium, *Phys. Rev.* **58**, 89.

Gauthier-Lafaye F., Holliger P. and Blanc P.-L. (1996) Natural fission reactors in the Franceville Basin, Gabon: A review of the conditions and results of a "critical event" in a geologic system, *Geochimica et Cosmochimica Acta* **60**, 4831–4852.

Gramaccioli C.M. and Segalstad T.V. (1978) A uranium- and thorium-rich monazite from a south-alpine pegmatite at Piona, Italy, *American Mineralogist* **63**, 757–761.

Groves L.R. (1964) *Now it Can be Told. The Story of the Manhattan Project* (Harper & Brothers, Publishers, New York).

Guardian Unlimited (2000) UN raises alarm on toxic risk in Kosovo, 22 March, www.guardian.co.uk/Archive/Article/0,4273,3976810,00.html.

Helean K.B., Burakov B.E., Anderson E.B., Strykanova E.E., Ushakov S.V. and Ewing, R.C. (1997) Mineralogical and microtextural characterization of "gel-zircon" from the Manibay Uranium Mine, Kazakhstan. *Mat. Res. Soc. Symp. Proc. Scientific Basis for Nuclear Waste Management XX*, Vol. **465**, 1219–1226.

Hoffman D.C. (ed.) (2002) *Advances in Plutonium Chemistry 1967–2000*, (American Nuclear Society, La Grand Park, Illinois, USA).

Hutton C.O. (1951) Uranium, thorite, and thorium monasite from black sand pay streaks, San Mateo County, Calif, *Geological Society of America Bulletin* **62**, 1518–1519.

Hyde E.K. and Seaborg G.T. (1957) *The Transuranium Elements*, Handbuch der physic, herausgegeben von S. Flugge, Band XLII, S. 205 (Springer-Verlag, Berlin-Gottingen-Heidelberg).

ICRP (1986) *The Metabolism of Compounds of Plutonium and Related Elements.* International Commission on Radiobiological Protection. Publication 48.

Il'in L.A. and Filatov V.A. (eds.) (1990) *Harmful Chemicals. Radioactive Materials*, Leningrad, "Himiya" (in Russian).

IRSN Final report (2001) Evaluation of nuclear criticality safety data and limits for actinides in transport, C4/TMR2001/200-1. IRSN. BP 17, 92262 FONTENAY-AUX-ROSES CEDEX, France. http://ec.europa.eu/energy/nuclear/transport/doc/irsn_sect03_146.pdf.

King S.J. and Putte D.V. (2003) Identification and description of UK radioactive wastes and materials potentially requiring long-term management, Nirex Report UK N/085, pp. 5–30.

Konev A.A., Ushyapovskaya Z.F., Kashaev A.A. and Lebedeva V.S. (1969) Tazheranite — New calcium-titanium-zirconium mineral, *Reports of USSR Academy of Sciences* **186**, 917–920 (in Russian).

54   *Crystalline Materials for Actinide Immobilisation*

Kosterin A.V. and Zuyev V.N. (1958) On the Zr/Hf ratio in some igneous rocks of North Kirgizia, *Geochimia* 1, 86–88 (in Russian).

KRI-75. *V.G. Khlopin Radium Institute — To the 75 Years Anniversary*, E.I. Ilyenko (ed.) (1997), St. Petersburg (in Russian).

Kumral M., Çoban H. and Caran Ş. (2007) Th-, U- and LREE-bearing grossular, chromian ferriallanite-(Ce) and chromian cerite-(Ce) in skarn xenoliths ejected from the Golcuk maar crater, Isparta, Anatolia, Turkey, *Canadian Miner.* 45, 1115–1129.

Lumpkin G.R., Smith K.L., Giere R. and Williams C.T. (2004) Geochemical behaviour of host phases for actinides and fission products in crystalline ceramic nuclear waste forms, In *Energy, Waste, and the Environment: A Geochemical Perspective*, eds. Giere R. and Stille P. (Geological Society, Special Publications, London), pp. 89–111,

Maas R., Kinny P.D., Williams I.S., Froude D.O. and Compston W. (1992) The Earth's oldest known crust: A geochronological and geochemical study of 3900–4200 Ma old detrital zircons from Mt. Narryer and Jack Hills, Western Australia, *Geochimica et Cosmochimica Acta* 56, 1281–1300.

Mark J.C. (1993) Explosive properties of reactor-grade plutonium, *Science and Global Security*, 4, 111–128.

Mitchell R.S. (1979) *Mineral Names — What Do They Mean?* (Van Nostrand Reinhold Company, New York).

Ojovan M.I. and Lee W.E. (2005) *An Introduction to Nuclear Waste Immobilisation* (Elsevier Science Publishers, Amsterdam), 315 pp.

Orlov V. (2004) Illegal traffic of nuclear materials, *IAEA Bulletin* 46, 63–65.

Patterson J.H., Nelson G.B. and Matlack G.M. (1974) Report LA-5624, Los Alamos National Laboratory, Los Alamos, NM.

Portnov A.M., Dubakina L.S. and Krivokoneva G.K. (1981) Murataite in predicted association with laundautite, *Reports of USSR Academy of Sciences* 261, 741–744 (in Russian).

Radioactive Waste Management Glossary (2003) 2003 Edition, IAEA, Vienna.

Seaborg G.T., McMillan E.M., Kennedy J.W. and Wahl A.C. (1946) Radioactive Element 94 from Neutrons on Uranium, *Phys. Rev.* 69, 366–367.

Simuya S. (1989) *Zircon — Science and Engineering* (Uchida Rokakuho Publishing Co., Ltd, Tokyo, Japan).

Smetannikov A.F. (1997) U-bearing zircon gels in U-Mo ores of the Glubinnoe deposit, Northern Kazakhstan, *Geochemistry International* 35, 487–490.

Smith J.V. (1974) Lunar mineralogy: A heavenly detective story presidential address, Part I. *Amer. Mineral.* 59, 231–243.

Soboleva M.V. and Pudovkina I.A. (1957) *Mineral of Uranium (Handbook)*, Moscow (in Russian).

Standards of Radiation Safety NRB-76/87 (1988) Energoatomizdat, Moscow (in Russian).

Standards of Radiation Safety (1999) Aprohim, Moscow (in Russian).

Stukin A.D. and Bystrova T.B. (2003) Plutonium. Production and isotope composition. Classification aspects, *Novosti FIS*, CNIIAtominform, 3, 5–7.

Van Wambeke L. (1970) The alteration processes of the complex tiano-niobo-tantalates and their consequences, *N. Jahrb. Mineral. Abh.* **112**, 117–149.

Vdovenko V.M. (1969) *Modern Radiochemistry* (Atomizdat, Moscow), (in Russian).

Yagovkina M. (2009) Private communication, Ioffe Physico-Technical Institute, St. Petersburg.

Zirlin V.A., Garbuzov V.M. and Kitsay A.A. (2008) Private Communication, V.G. Khlopin Radium Institute.

Zubkov L.B. (1989) *Metal of Gold-colored Stone*, (Nauka, Moscow), (in Russian).

Chapter 2

Current and Potential Actinide Applications

2.1. Advanced Nuclear Fuel Cycle

Most current commercial nuclear reactors such as the pressurised
water reactor (PWR), boiling water reactor (BWR) and advanced
gas cooled reactor (AGR) are thermal reactors and use uranium as
fuel in the form of UO_2 ceramic, in which the main burnable component is ^{235}U.

2.1.1. *MOX nuclear fuel*

Development of mixed oxide (MOX) nuclear fuel was initiated by the
need to immobilise excess civilian and weapons-grade plutonium. In
contrast to uranium-oxide fuel, MOX nuclear fuel contains plutonium and consists of two phases: UO_2 and PuO_2; and/or a single-
phase solid solution $(U,Pu)O_2$. The PuO_2 content may vary from 1.5
wt.% to 25–30 wt.% depending on the type of reactor. Since 1972,
world production of plutonium has exceeded demand for all purposes
including research, and since 1975, all developed nuclear countries
have accumulated plutonium as a prospective material for nuclear
energy (Kotelnikov *et al.*, 1978). Although MOX fuel may be used in
thermal reactors to provide energy, efficient burning of plutonium in
MOX can only be achieved in fast reactors. The content of unburnt

plutonium in spent MOX fuel from thermal reactors is significant —
greater than 50% of the initial plutonium loading. However, during
the burning of MOX the ratio of fissile (odd numbered) isotopes to
non-fissile (even) drops from around 65% to 20%, depending on burn-
up. This makes any attempt to recover the fissile isotopes difficult
and any bulk Pu recovered would require such a high percentage of
Pu in any second generation MOX that it would be impractical. This
means that such a spent fuel would be hard to reprocess for further
reuse (burning) of plutonium. Regular reprocessing of biphasic spent
MOX is difficult because of the low solubility of PuO_2 in nitric acid.
Although the $(U,Pu)O_2$ solid solution dissolves more slowly than
pure UO_2, its solubility is higher than that of PuO_2, therefore single-
phase MOX fuel based on a solid solution of $(U,Pu)O_2$ is expected to
be easier to reprocess. However, the reprocessing technology needs
to be proven using irradiated samples. Formation of $(U,Pu)O_2$ solid
solution can take place in the matrix of biphasic MOX at high temperature ($>1,400°C$) under irradiation (Freshley, 1973).

It is important to note that MOX irradiation is accompanied by
redistribution of plutonium in the fuel matrix with the highest Pu
content observed at the fuel pellet centres. The gradient of plutonium
content in different areas of fuel pellets can achieve 7–10% (Chikalla
et al., 1964, Sari *et al.*, 1970). Inhomogeneous distribution of plutonium arising in MOX during irradiation (at high burn-up) may
affect the mechanical durability of the fuel cladding as a result of
local swelling of the fuel matrix.

2.1.2. *Ceramic nuclear fuel*

Development of "ceramic" (also termed "rock-like" or "inert") fuel
is under consideration in many countries. This fuel has the following
ideal goals:

1. Increased fuel burn-up (maximum burning of U and Pu);
2. Possible transmutation of minor actinides (Np, Am, Cm) accompanying fuel irradiation;
3. Environmental safety of spent ceramic fuel, which would be
   considered as the final wasteform.

The first two can be successfully achieved, separately or together, without using durable actinide host phases. In this case the oxides of U and (or) Pu in a ceramic matrix are just mixed with particles of radiation-resistant and chemically-inert phases such as CeO_2, MgO, Y_2O_3, Al_2O_3, $MgAl_2O_4$, $BaZrO_3$ and $Y_3Al_5O_{12}$ in so-called Inert Matrix Fuels (IMFs). The role of the non-radioactive inert phases is mainly to confer stable mechanical behaviour (low matrix swelling, crack resistance and limited crack interaction with cladding) on the nuclear fuel at high burn-up and high temperature, and on accumulation of gaseous fission products. Preliminary irradiation experiments have demonstrated stable behaviour of three fuel compositions based on mixtures of UO_2 with CeO_{2-x}, MgO and Y_2O_3 but not with $Y_3Al_5O_{12}$ (Neeft *et al.*, 2003). Inert matrices of such spent ceramic fuel might be considered a mechanical barrier to further radionuclide migration in the environment but the occurrence of separate actinide phases in these ceramics does not yet justify allowing them to be used directly as wasteforms. Suitable ceramic fuel might be obtained, however, if all actinides are incorporated into the crystalline structure of durable host phases as solid solutions: for example, $(Zr,Y,Pu)O_2$, $(Pu,Zr)N$ and $(Th,Zr,U,Pu)O_2$. Cubic zirconia incorporating plutonium (and other actinides) remains the most likely material for ceramic fuel (Carroll, 1963). Burn-up calculations carried out for dual-phase ceramic fuel based on zirconia-plutonia solid solution $(Zr,Y,Gd,Pu)O_2$ and Al_2O_3 have demonstrated the possibility of burning >80% of the total plutonium and almost all [239]Pu in a PWR (Furuya *et al.*, 1996).

2.1.3. *Advanced nuclear reactors*

The main disadvantages of modern nuclear reactors using U-oxide or MOX fuel are:

1. Low burn-up of [235]U and [239]Pu;
2. The spent nuclear fuel matrix containing long-lived radionuclides is not chemically durable or stable in a likely geological repository environment;

3. Direct geological disposal of spent fuel as a wasteform is difficult
   to justify. Burying huge volumes of ^{238}U and fissile ^{235}U and ^{239}Pu
   does not follow the requirements of environmental safety and proliferation resistance. In addition, it is not sensible economically;
4. Reprocessing of spent fuel is an expensive and waste generating
   process. In some developed nuclear countries reprocessing is not
   currently used (e.g., USA, Sweden).

Therefore, new generations of nuclear reactors must be developed to
solve the problem of maximum fissile material burn-up. In the near
future this is likely to be by using fast reactors which induce greater
transmutation of plutonium and other actinides. Another possibility
is the development of reactors that can use new types of fuel such as:

1. Ceramic fuel for high burn-up followed by direct geological disposal of spent fuel (without reprocessing);
2. Thorium fuel (^{232}Th-^{233}U or ^{232}Th-^{233}U-^{239}Pu) which accumulates smaller amounts of minor actinides. Development of Th as
   an additional source of natural nuclear material is well advanced
   in India.

2.2. Inert Pu Ceramic Fuel

The use of inert ceramic fuel in thermal or fast reactors is a promising way to give maximum plutonium burn-up followed by direct
disposal of spent fuel (see Sec. 2.1.2). Ceramic Pu fuel is likely to
be more competitive than MOX, which provides sufficient plutonium burn-up only in fast reactors. Furthermore, after irradiation
MOX is not a chemically-durable wasteform. Although such ceramic
fuels have been discussed for many years — for example, based on
(Zr,Pu)O_2 cubic solid solution (Carroll, 1963), and later cubic zirconia, was proposed as a candidate wasteform for actinides (Heimann
and Vandergraaf, 1988) — the development of such fuel is still at a
preliminary stage. Under irradiation in a high-flux research reactor
the behaviour of single-phase fuel based on (Er,Y,Pu,Zr)O_{2-x} solid

solution was comparable to MOX (Degueldre and Hellwig, 2003). Stable behaviour of multi-phase fuel based on $(Zr,Y,Pu)O_2$ solid solution (11.1. wt.% PuO_2), $MgAl_2O_4$ and Al_2O_3 was demonstrated at burn-up 184, 192 and 237 GWd/m^3 (Yamashita *et al.*, 2003). However, the thermal conductivity of zirconia-based ceramic fuel is less than half that of UO_2. This can be compensated for either by making fuel pellets with a special shape or using additives such as metals (e.g., Mo or Zr) or other highly-conducting but inert phases. It is important to emphasise the economic advantage of using zirconia-based plutonium fuel. Recycled zircaloy cladding from spent oxide fuel is an ideal source of zirconium to be used for fabrication of zirconia-based ceramic fuels or wasteforms (Burakov and Anderson, 1998).

2.3. Sealed Radioactive Sources

Sealed radioactive sources (SRS) are used extensively, e.g., in hospitals for medical applications. The worldwide inventory of SRS is large. For example, in the European Union about 500,000 SRS have been sold, most of which remain in use or in store (Angus *et al.*, 2000). Many of these sources are characterised by very high levels of radioactivity. The decay of radionuclides in SRS diminishes their high levels of radioactivity insufficiently for subsequent safe handling and disposal. Thus, most SRS continue to present a high radiological hazard beyond their design lifetime and need to be managed and disposed of safely to ensure the long-term protection of people and the environment. As the radiotoxicity of SRS is high, they require special and expensive procedures for safe and secure storage, and most highly active and long-lived SRS remain in store pending the availability of a suitable disposal option (Chapman *et al.*, 2005). Actinide sealed radioactive sources are numerous and their compositions well known (Sytin *et al.*, 1984; Alardin *et al.*, 2003; Ojovan and Lee, 2005). In this chapter we summarise briefly the typical features of actinide-containing SRS. Table 2.3.1 gives their most frequent uses.

The most common devices using SRS with relatively small quantities of ^{241}Am, ^{238}Pu and ^{239}Pu are smoke detectors (Fig. 2.3.1). These

Table 2.3.1.   Summary of the principal parameters of radionuclides used in actinide-containing SRS (modified from Alardin *et al.*, 2003).

Radionuclide	Half-life, y	Emission	Energies, MeV	Application	Activity range, Bq
^{226}Ra	1,600	α, γ	4.8 (α), 0.186 (γ)	Brachytherapy	$3 \ 10^6 - 3 \ 10^7$
^{226}Ra/Be	1,600	neutron		Well logging, moisture detector	
^{237}Np	$2.2 \ 10^6$	α, γ	4.8 (α), 0.029 (γ)	Measurement of neutron fluxes	
^{238}Pu	87.7	α, γ	5.5 (α), 0.044 (γ)	Static electricity eliminator	Up to $3 \ 10^8$
				Thermoelectric generator	Up to $1.9 \ 10^{12}$
^{238}Pu				X-ray fluorescence analyser	$6.3 \ 10^8 - 3.7 \ 10^{10}$
				Well logging	$1.2 \ 10^{10} - 1.2 \ 10^{12}$
				Moisture detector	
^{239}Pu	24,181	α, γ	5.2 (α), 0.052 (γ)	Smoke detectors, gas analysers, gas chromatography	$\sim 10^4$
^{241}Am	432.2	α, γ	5.5 (α), 0.060 (γ)	Density measurement	$10^9 - 4 \ 10^9$
				Static electricity eliminator	$10^9 - 1.2 \ 10^{12}$
				Smoke detectors	$2 \ 10^4 - 3 \ 10^6$
				Lightning arresters	$5 \ 10^7 - 5 \ 10^8$
				Bone densitometry	$10^9 - 10^{10}$
				X-ray fluorescence analyser	$3.7 \ 10^8 - 1.85 \ 10^{11}$
^{241}Am/Be		neutron		Well logging	$10^9 - 8 \ 10^{11}$
				Moisture detector	$10^9 - 9.25 \ 10^9$

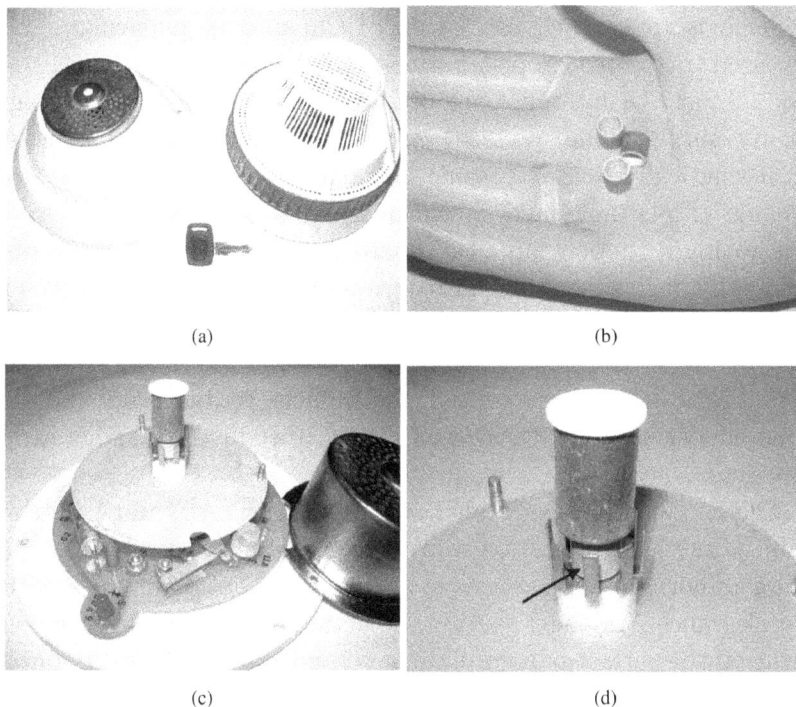

Fig. 2.3.1. Plutonium-sealed sources in old Russian smoke detectors. (a) Smoke detectors RID-6m and IDF-1m; (b) Three Pu sealed sources AIP-EDGH used in smoke detectors. Plutonium isotopes (18.5 MBq of ^{239}Pu or ^{238}Pu) are fixed by a thin layer of TiO_2 on an external surface of alumina cylinder; (c) General view of a dismantled smoke detector RID-6m; (d) Detailed image of device with Pu sealed source (arrow). Courtesy of V. Zirlin, the V.G. Khlopin Radium Institute.

sealed sources are made of ceramics, however, neither americium nor plutonium are incorporated into the crystalline structure of durable host phases. During source fabrication the actinide-containing solution is evaporated on the surface of some ceramics (for example, Al_2O_3) and the actinide is then fixed by sintering at 1,200–1,400°C with a small amount of borosilicate glass or by precipitation of a thin layer (several micrometres) of TiO_2. New safety requirements identify smoke detectors with actinide sealed sources as a subject of serious concern. More recent detectors use advanced non-radioactive analogues.

64   *Crystalline Materials for Actinide Immobilisation*

Some isotopes such as ^{242}Cm; ^{244}Cm and in particular ^{238}Pu are used in radioisotope thermoelectric generators (RTG) (see Sec. 1.3.2). The internal core of these sources consists of a high-density sintered (hot-pressed) actinide oxide matrix. The external hermetically-welded capsule of modern electric-heat generators is made of platinoid alloys, which are chemically inert and resistant to high temperatures (at least 1,500°C). However, this does not justify using Pu and Cm oxides as the matrices for SRS. The main disadvantage of actinide oxide matrices was revealed in April 1964 by an accident with the US space vehicle carrying the ^{238}Pu RTG SNAP-9A, which caused the vehicle to burn up and disperse almost 1 kg of ^{238}Pu (0.63 TBq ^{238}Pu) in the Earth's atmosphere.

The largest group of sealed sources with ^{241}Am and ^{238}Pu are gamma and X-ray radiation sources for radiometric analysis in metallurgy, fluorographic analysis of alloys, control of boundaries, levels and clearances. In these sources the actinide is usually fixed by sintering on the surface of a graphite or ceramic matrix. In some sources the core is made of a pressed mixture of Am-oxide with metallic aluminium powder, which is sealed in a welded titanium or stainless steel capsule (Fig. 2.3.2). In X-ray radiation sources the metal capsule is

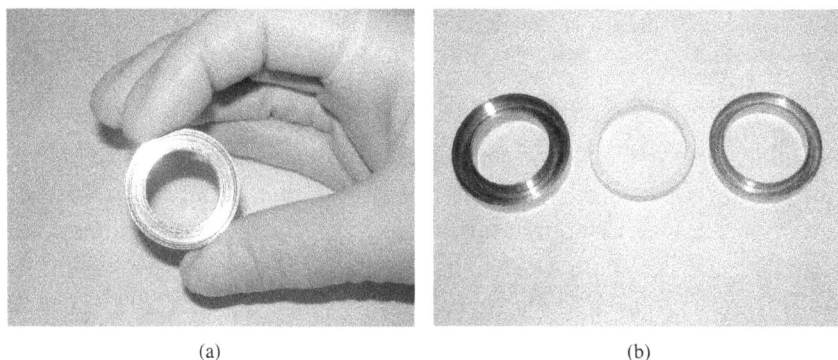

(a)   (b)

Fig. 2.3.2.   Americium ring sealed source GAm1.032 (1,850 or 3,700 MBq ^{241}Am) produced by RITVERC Isotope Products, Russia; (a) General view; (b) three parts of dismantled source with the translucent ceramic core between stainless steel body and lid. Courtesy of V. Zirlin, the V.G. Khlopin Radium Institute.

Fig. 2.3.3.   Plutonium sealed source AIP-N (produced by Association Production MAYAK, Russia) used as a static electricity eliminator. Pu isotopes (37 MBq of ^{239}Pu or ^{238}Pu) are fixed in an enamel layer on a steel plate. Courtesy of V. Zirlin, the V.G. Khlopin Radium Institute.

supplied with a brazed beryllium window with the Be foil thickness not exceeding 50 micrometres.

Sealed Pu sources used to eliminate electrostatic charge are characterised by an unusual design (Fig. 2.3.3). Their important feature is the absence of an external capsule. Plutonium is fixed (dissolved) in the vitreous enamel layer covering a steel support.

Typical SRS lifetimes are less than 15 years so they accumulate relatively quickly as waste.

2.4. Self-glowing Materials

Development of self-glowing actinide-doped materials with durable crystalline matrices is a new area of research that may significantly change the approaches to actinide immobilisation. Addition of alpha-emitting radionuclides to some glasses and crystals or external alpha-irradiation of some solids may confer luminescence (Bagnall and D'Eye, 1954). This phenomenon was utilised in glowing ^{226}Ra-containing paints in the early 20th century although their poor

66   *Crystalline Materials for Actinide Immobilisation*

chemical durability and weak radiation resistance mean this practice has discontinued. Synthesis and study of non-radioactive scintillating materials, however, continue to increase. The same solids
may scintillate under UV, X-ray, gamma, beta and alpha irradiation
with the intensity of luminescence strongly dependent on the energy
of excitation. For example, mono-crystalline zircon, $(Zr,Pu)SiO_4$,
doped with 2.4–2.7 wt.% ^{238}Pu immediately after synthesis revealed
weak glowing from Pu self-irradiation (Fig. 2.4.1(a)), low intensity
of cathodoluminescence and no UV-luminescence (Hanchar *et al.*,
2003). In general, scintillation is caused by luminescence centres,
which are produced by excitation of admixture activating elements
such as Eu, Ce, Tb, Mn, Gd, Nd, Er, Yb, Dy or intrinsic defects.
The chemical element directly responsible (as activating element) or
indirectly (causing intrinsic defects) for luminescence is called the
"phosphor". Crystalline scintillating materials with high chemical
resistance, mechanical durability, and stability under irradiation are
promising for long-term operation in optical couplers, robotics and
medicine.

Such materials could last tens to hundreds of years in aggressive
chemical media and in space. The intensive scintillation might also be
converted into electric current. This opens up the possibility of developing long-lasting nuclear electric batteries using stable materials
glowing under irradiation from different radionuclides: in particular,
alpha-emitters (Mikhalchenko, 1988; Sychov *et al.*, 2008). Obviously,

Table 2.4.1   Features of some ^{238}Pu-doped self-glowing crystals synthesised
at the V.G. Khlopin Radium Institute (Burakov *et al.*, 2009).

Crystal	Phosphor	^{238}Pu content, wt.%	Self-glowing intensity
Zircon, $(Zr,Pu,...)SiO_4$	None	2.4–2.7	Weak
	Eu	0.007	Weak
	In	0.02	Weak
	In	0.01	High
	In + Tb	0.01	Very high
Xenotime, $(Y,Pu,Eu)PO_4$	Eu	0.002	Weak
	Eu	0.1	Very high

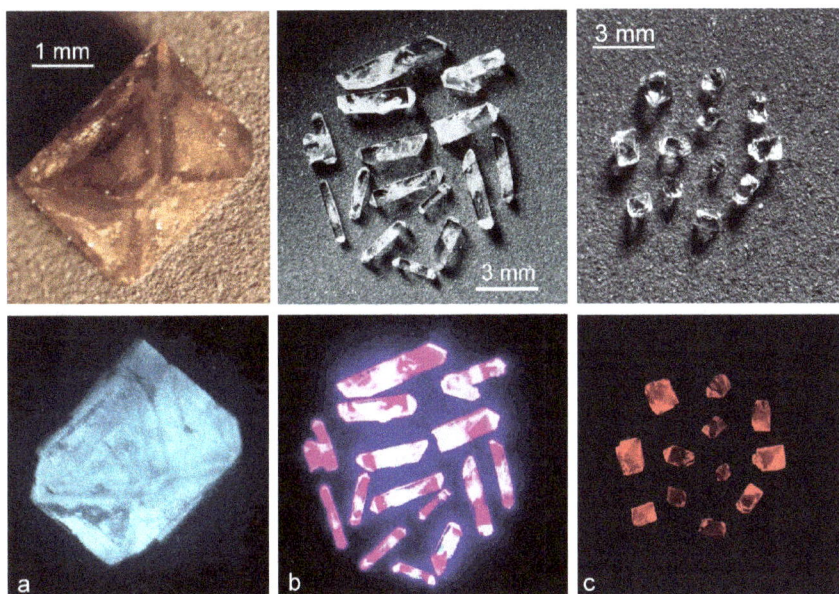

Fig. 2.4.1.   Self-glowing crystals in transmitted light (top) and in the dark (lower row): (a) phosphor-free zircon containing 2.4 wt.% ^{238}Pu; (b) zircon doped with 0.007 wt.% ^{238}Pu and Eu; (c) xenotime doped 0.1 wt.% ^{238}Pu and Eu (Burakov *et al.*, 2009).

low-radioactivity devices are the most likely for industrial-scale application. Disposal of spent nuclear batteries could be ecologically safer if the radionuclide is immobilised in the crystalline structure of a durable scintillating host phase. The safest self-glowing materials could be those based on durable crystalline host phases accommodating a relatively small amount of radionuclides (causing excitation and luminescence) and an optimal amount of phosphor (responsible for the highest intensity of luminescence). The best phosphor for a selected durable host phase must first be identified and following this, a balance between phosphor(s) and radionuclide contents in the host phase (Table 2.4.1) must be achieved.

Both excessive and insufficient phosphor and/or radionuclide contents suppress the self-glowing intensity. It is desirable to keep the radionuclide content as low as possible to avoid radiation damage effects and to meet safety requirements. Experiments on synthesis of self-glowing materials doped with radionuclides and particularly actinides are expensive. It is reasonable to identify the optimal content of phosphor before doping the same host phase with actinides. Cathodoluminescence (CL) spectroscopy of non-radioactive solids could be used as an important tool for such a purpose (Burakov *et al.*, 2007). Crystals of selected host phase with various phosphor contents have been studied under an electron beam causing CL (see Sec. 5.3. Cathodoluminescence). Although the energy of the electron beam is not equal to that of alpha-radiation, the highest CL intensity roughly correlates with the optimal phosphor content of an actinide-doped self-glowing analogue of the same host phase (Burakov *et al.*, 2009). Some self-glowing crystals obtained in the V.G. Khlopin Radium Institute are shown in Fig. 2.4.1. The self-glowing intensity for both phosphor-free zircon containing 2.4–2.7 wt.% ^{238}Pu (Fig. 2.4.1(a)) and Eu-doped zircon containing only 0.007 wt.% ^{238}Pu (Fig. 2.4.1(b)) was at the same weak level. CL study of zircon and xenotime (Fig. 2.4.1(c)) allowed appropriate phosphors to be selected (Eu for xenotime and In + Tb for zircon) and identification of optimal phosphor contents for synthesis of crystals with intensive self-glowing at constant ^{238}Pu content (Table 2.4.1).

2.5. Transmutation Targets

Ceramic targets containing actinides that can be bombarded with neutrons to induce transmutation reactions to remove the most difficult long-lived species are similar to IMFs (see Sec. 2.1.2) consisting of actinide-containing solid solutions such as (Am,Zr)N, (Am,Y)N, (Zr,Cm)O$_2$, (Zr,Cm,Am)O$_2$, (Zr,Am,Y)O$_2$ or just actinide phases like AmO$_2$, NpO$_2$, NpN, AmN mixed with some inert phases such as MgO, MgAl$_2$O$_4$, (Zr,Y)O$_2$, TiN and ZrN. The role of the non-radioactive inert phases is mainly to provide stable mechanical behaviour to the target under neutron irradiation. Crystalline (Zr,Cm,Am)O$_2$ solid solutions with fluorite and pyrochlore structures have been successfully synthesised and studied as transmutation targets (Raison and Haire, 2003). Synthesis of nitride solid solutions (Am$_{0.1}$Zr$_{0.9}$)N and (Pu$_{0.1}$Zr$_{0.9}$)N for irradiation tests has also been reported (Minato *et al.*, 2003). Transmutation targets containing spherical inclusions of actinide-doped cubic zirconia, $(Zr,Y,Am)O_2$ and $(Zr,Y,Pu)O_2$, in an MgO matrix have been fabricated (Croix-marie *et al.*, 2003). Under irradiation inert phases in transmutation targets may react with actinide phases. Hibonite, $MgPuAl_{11}O_{19}$, may have arisen from Am transmutation in a high thermal flux reactor (Wiss *et al.*, 2003). The initial target was an $MgAl_2O_4$ matrix with inclusions of Am-oxide and possibly, $AmAlO_3$.

2.6. Summary

Using actinides in nuclear fuel, sealed radioactive sources or advanced materials such as self-glowing crystals has many potential benefits. However, a serious concern is the extremely high radiotoxicity of actinides and their migration in the environment (Ojovan and Lee, 2005). The use of chemically unstable forms of actinides in MOX and sealed radioactive sources is not appropriate by modern safety standards. There is a challenge to develop stable and durable actinide-bearing materials, which provide safe storage, use and final disposal. A key need is application of actinide solid solutions in durable crystalline host phases.

References

Alardin M., Al-Moughrabi M., Beer H.F., Grimm J., Menuhr H., Neubauer J., Niels Y., Ojovan M., Prasil Z., Shaddad I., Smith M., Tsyplenkov V. and Wells D. (2003) Management of disused long-lived sealed radioactive sources (LLSRS), IAEA-TECDOC-1357, Vienna, IAEA.

Angus M.J., Crumpton C., McHugh G., Moreton A.D. and Roberts P.T. (2000) Management and disposal of disused sealed sources in the European Union, European Commission, Report EUR 18186 EN, EC, Luxemburg.

Bagnall K.W. and D'Eye R.W.M. (1954) The preparation of polonium metal and polonium dioxide, *J. Chem. Soc. Lon.* 4295–4299.

Burakov B.E. and Anderson E.B. (1998). Development of crystalline ceramic for immobilization of TRU wastes in V.G. Khlopin Radium Institute, *Proc. Int. Conf. NUCEF'98, 16–17/11/98*, JAERI-Conf.99-004 (Part I), Hitachinaka, Ibaraki, Japan, 295–306.

Burakov B.E., Garbuzov V.M., Kitsay A.A., Zirlin V.A., Petrova M.A., Domracheva Ya.V., Zamoryanskaya M.V., Kolesnikova E.V., Yagovkina M.A. and

Orlova M.P. (2007) The use of cathodoluminescence for the development of durable self-glowing crystals based on solid solutions YPO_4-$EuPO_4$, *Semiconductors* **41**, 427–430.

Burakov B.E., Domracheva Ya.V., Zamoryanskaya M.V., Petrova M.A., Garbuzov V.M., Kitsay A.A. and Zirlin V.A. (2009) Development and synthesis of durable self-glowing crystals doped with plutonium, *J. Nucl. Mater.* **385**, 134–136.

Carroll D. (1963) The system PuO_2-ZrO_2, *J. Am. Ceram. Soc.* **46**, 194–195.

Chapman N., Cochran J., Davis P., Dayal R., Gera F., Harries J., Martianov V., Nel L., Ormai P., Ozhovan M. and Pla E. (2005) Disposal options for disused radioactive sources, TRS 436, 51p, IAEA, Vienna.

Chikalla T.D., McNeilly C.E. and Skavdahl R.E. (1964) Plutonium-oxygen system, *J. Nucl. Mater.* **12**, 131–134.

Croixmarie Y., Abonneau E., Fernandez A., Konings R.J.M., Desmouliere F. and Donnet L. (2003) Fabrication of transmutation fuels and targets: The ECRIX and CAMIX-COCHIX experience, *J. Nucl. Mater.* **320**, 11–17.

Degueldre C. and Hellwig, C.H. (2003) Study of a zirconia based inert matrix fuel under irradiation, *J. Nucl. Mater.* **320**, 96–105.

Freshley M.D. (1973) UO_2-PuO_2: A demonstrated fuel for plutonium utilization in thermal reactors, *Nucl. Tech.* **18**, 141–170.

Furuya H., Muraoka S. and Muromura T. (1996) Feasibility of rock-like fuel and glass waste form for disposal of weapons plutonium, in *Disposal of Weapon Plutonium*, eds. Merz E.R. and Walter C.E., NATO ASI series (Kluwer Academic Publisher, Dordrecht), 107–121.

Hanchar J.M., Burakov B.E., Anderson E.B. and Zamoryanskaya M.V. (2003) Investigation of single crystal zircon, $(Zr,Pu)SiO_4$, doped with [238]Pu and [239]Pu, in *Scientific Basis for Nuclear Waste Management XXVI*, eds. Finch R.J. and Bullen D.B., *Mat. Res. Soc. Symp. Proc.* **757**, 215–225.

Heimann R.B. and Vandergraaf T.T. (1988) Cubic zirconia as a candidate waste form for actinides: Dissolution studies, *J. Mater. Sci. Lett.* **7**, 583–586.

Kotelnikov P.B., Bashlikov S.N., Kashtanov A.I. and Melnikova T.S. (1978) *High Temperature Nuclear Fuel* (Atomizdat, Moscow), in Russian.

Mikhalchenko G.A. (1988) *Radioluminescence Illuminators* (Energoatomizdat, Moscow), in Russian.

Minato K., Akabori M., Takano M., Arai Y., Nakajima K., Itoh A. and Ogawa T. (2003) Fabrication of nitride fuels for transmutation of minor actinides, *J. Nucl. Mater.* **320**, 18–24.

Neeft E.A.C., Bakker K., Schram R.P.C., Conrad R. and Konings R.J.M. (2003) The EFTTRA-T3 irradiation experiment on inert matrix fuels, *J. Nucl. Mater.* **320**, 106–116.

Ojovan M.I. and Lee W.E. (2005) *An Introduction to Nuclear Waste Immobilisation* (Elsevier Science Publishers B.V., Amsterdam).

Raison P.E. and Haire R.G. (2003) Structural investigation of the pseudo-ternary system AmO_2-CmO_2-ZrO_2 as potential materials for transmutation, *J. Nucl. Mater.* **320**, 31–35.

Sari C., Benedict U. and Blan H. (1970) A study of the ternary system UO_2-PuO_2-Pu_2O_3, *J. Nucl. Mater.* **35**, 267–277.

Sychov M., Kavetsky A., Yakubova G., Walter G., Yousaf S., Lin Q., Chan D., Socarras H. and Bower K. (2008) Alpha indirect conversion radioisotope power source, *Appl. Radiat. Isot.* **66**, 173–177.

Sytin B.P., Teplov F.P. and Cherevatenko G.A. (1984) *Radioactive Sources of Ionizing Radiation* (Energoatomizdat, Moscow), in Russian.

Wiss T., Konings R.G.M., Walker C.T. and Thiele H. (2003) Microstructure characterisation of irradiated Am-containing $MgAl_2O_4$, *J. Nucl. Mater.* **320**, 85–95.

Yamashita T., Kuramoto K., Nitani N., Nakano Y., Akie H., Nagashima H., Kimura Y. and Ohmichi T. (2003) Irradiation behavior of rock-like oxide fuels, *J. Nucl. Mater.* **320**, 126–132.

Chapter 3

Waste Actinide Immobilisation

3.1. Ceramic Nuclear Wasteforms: Historical Overview

3.1.1. *Early work*

Briefly summarised here is recent information on durable actinide host phases including those ceramics initially not considered as wasteforms. Detailed overviews of ceramic wasteforms such as Synroc and tailored ceramics have been published (Lutze and Ewing, 1988; Donald *et al.*, 1997; Ewing, 1999) while other nuclear wasteforms including glasses and glass composite materials have also been reviewed recently (Stefanovsky *et al.*, 2004; Yudintsev *et al.*, 2007; Ojovan and Lee, 2007; Caurant *et al.*, 2009). The first proposed use of minerals as nuclear wasteforms was made by Hatch (Hatch, 1953). Carroll then proposed using zirconia-plutonia solid solution, $(Zr,Pu)O_2$, as an advanced nuclear fuel (Carroll, 1963). Absorption spectroscopy results for U^{4+} in single-crystal zircon grown by a flux method were published in 1967 (Richman *et al.*, 1967) although no information on the uranium content in the zircon crystals was provided at that time. Early work at Pacific Northwest National Laboratory (PNNL) and Pennsylvania State University (PSU) was performed in the 1970s on Supercalcine, regarded as the forerunner of Synroc (IAEA 257, 1985) with the aim of achieving a waste loading

74   *Crystalline Materials for Actinide Immobilisation*

of 60% by additions to HLW liquid which on calcination and sintering at <1100°C formed durable crystalline phases (McCarthy, 1979).
The idea of "tailor-made" glass-ceramics was also born at PSU (Roy,
1975) and was based on using chemical additives to induce crystallisation of specific durable phases in glass wasteforms for immobilisation
of HLW. The first tailored titanate ceramics were synthesised by hot
pressing at Sandia Labs, USA (Lutze and Ewing, 1988). Prospective
phases for actinide accommodation such as zirconolite, $CaZrTi_2O_7$,
and perovskite, $CaTiO_3$, were identified in this work. Further development of tailored ceramics for immobilisation of HLW between
1976 and 1982 identified other prospective actinide host phases such
as: Ti-pyrochlore, $(Gd,La)_2Ti_2O_7$; cubic fluorite-type solid solutions,
$(U,Th,Zr)O_2$, and murataite-type phase, $Zr(Ca,Mn)_2(Fe,Al)_4Ti_3O_{16}$
(Morgan and Ryerson, 1982). Zirconolite $CaZrTi_2O_7$, and perovskite,
$CaTiO_3$, are well-known actinide host phases in the polyphase matrix
of the ceramic wasteform Synroc developed in Australia (Ringwood,
1978). A ceramic wasteform based on monazite, $(Ln,Gd)PO_4$, was
proposed in 1978 (Boatner, 1978; McCarthy *et al.*, 1978). Researchers
at PNNL, USA reported that the orthosilicate apatite phase can
incorporate at least 6 wt.% actinides ([244]Cm and [240]Pu) in a devitrified waste glass (Weber *et al.*, 1979). Single crystals of monazite,
doped with [239]Pu (5.0 wt.%); [237]Np (2.0 wt.%),[241]Am (0.2 wt.%);
[246]Cm (0.2 wt.%) were obtained at Oak Ridge National Laboratory (ORNL), USA (Boatner *et al.*, 1980). [243]Cm and [244]Cm doped
zircon-structured $LuPO_4$ single crystals have been grown at ORNL
(Abraham and Boatner, 1982). Workers at Los Alamos National Laboratory (LANL), USA, synthesised [238]Pu-substituted cubic zirconolite, $CaPuTi_2O_7$, and partially substituted monoclinic zirconolite,
$CaZr_{0.8}Pu_{0.2}Ti_2O_7$ (Clinard *et al.*, 1982).

Zircon, $ZrSiO_4$, as a suitable wasteform was mentioned for the
first time in 1982 (McKown *et al.*, 1982). Silicate apatite,
$Ca_2Nd_8(SiO_4)_6O_2$, doped with 2.3 mol.% Cm_2O_3 (62% [244]Cm) was
synthesised and studied at PNNL (Weber, 1982 and 1983). Workers at PNNL synthesised polycrystalline zircon, $(Zr,Pu)SiO_4$, doped
with 10 wt.% [238]Pu (Exharos, 1984), Cm-doped (3 wt.% [244]Cm)
pyrochlore, $Gd_2Ti_2O_7$, and zirconolite, $CaZrTi_2O_7$ (Weber *et al.*,

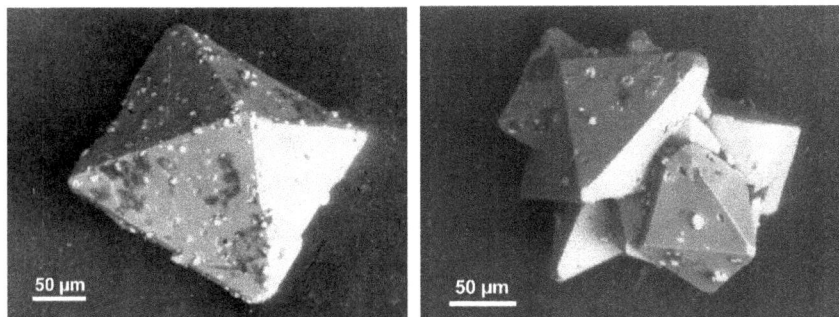

Fig. 3.1.1.   Crystals of high-uranium zircon, $(Zr,U)SiO_4$, extracted from the matrix of the Chernobyl "lava". The range of zircon's chemical composition is within the formula interval $(Zr_{0.95}U_{0.05})SiO_4$–$(Zr_{0.90}U_{0.10})SiO_4$ (Burakov *et al.*, 1991; Burakov *et al.*, 1993; Anderson *et al.*, 1993).

1986). Synroc doped with ^{238}Pu and ^{244}Cm was prepared in the UKAEA Laboratories at Harwell, UK in 1987 (Hambley *et al.*, 2008).

Cubic zirconia was proposed as a prospective actinide wasteform in 1988 (Heimann and Vandergraaf, 1988). Zircon single crystals doped with 0.2 and 2.0 wt.% ^{237}Np and separately 0.2 wt.% ^{242}Pu were studied optically in 1988–1989 (Poirot *et al.*, 1988; Poirot *et al.*, 1989). Crystals of synthetic high-uranium zircon, $(Zr,U)SiO_4$, containing up to 10 wt.% U in the form of solid solution were discovered in the matrices of Chernobyl "lava" (Fig. 3.1.1) by researchers of the V.G. Khlopin Radium Institute (KRI), Russia (Burakov *et al.*, 1991). Together with high-uranium zircon inclusions of monoclinic zirconia, $(Zr,U)O_2$, containing several wt.% U in solid solution were identified (Fig. 3.1.2). Since then zircon and zirconia have been considered to be promising host phases for immobilisation of actinides (Burakov, 1993; Anderson *et al.*, 1993).

3.1.2.   *Emergence of Pu wasteforms*

In 1990 Russian programmes on development of ceramic wasteforms for HLW immobilisation were initiated by the Ministry for Atomic Energy. In contrast to Australian Synroc, which was developed originally for immobilisation of bulk HLW from spent fuel reprocessing,

Fig. 3.1.2.   Backscattered electron SEM image of inclusions of U-doped zirconia, $(Zr,U)O_2$, in crystal of high-uranium zircon, $(Zr,U)SiO_4$. Black matrix is silicate glass-like material of Chernobyl "lava". White inclusions are crystals of U-oxide with Zr admixture (Anderson *et al.*, 1993).

the main goal of the Russian ceramics research was to immobilise separately the fraction of transplutonium elements and radioactive rare-earths after partitioning of HLW. Different prospective host phases were considered including zircon, zirconia, monazite, perovskite and zirconolite. Zircon and zirconia were identified as favorable host phases taking into account that zircaloy cladding of spent fuel is a significant source of zirconium.

The US Academy of Sciences stated in 1994 that excess plutonium is a danger to national and international security (Management and Disposition of Excess Weapons Plutonium, 1994). Since that time there have been intensive research efforts to develop wasteforms for plutonium immobilisation in the USA. A titanate ceramic based on pyrochlore, $(Ca,Gd,Pu,U,Hf)_2Ti_2O_7$, was developed by Lawrence Livermore National Laboratory (LLNL), USA, in co-operation with Australian Nuclear Science and Technology Organization (ANSTO) (Ebbinghaus *et al.*, 1998).

A detailed justification of zircon as a durable host phase for plutonium immobilisation was published in 1995 (Ewing *et al.*, 1995).

A uranium-free rock-like fuel based on cubic zirconia-plutonia solid solution was suggested for plutonium incineration in LWR or PWR (Degueldre *et al.*, 1996; Furuya *et al.*, 1996). Samples of poly-crystalline ^{237}Np doped yttria-stabilised cubic zirconia were syn-thesised at the Japan Atomic Energy Research Institute (JAERI) (Kuramoto *et al.*, 1995).

Rare-earth   silicate   apatite, $Ca_{4-x}REE_{6+x}(SiO_4)_{6-y}(PO_4)_y$ $(F,OH,O)_2$,was discussed as a prospective wasteform for dis-posal of plutonium (Ewing *et al.*, 1996). A murataite-like phase, $(Ca_{2.65}U_{0.3}Ce_{0.2})$ $Ti_{7.3}Mn_{0.6}Zr_{0.4}Al_{0.3}O_{20.0}$, was identified in the matrix of inductively-melted Synroc obtained in SIA "Radon", Rus-sia (Sobolev *et al.*, 1997). High resistance of cubic zirconia to low and high energy Xe ion irradiation damage was found (Deguel-dre *et al.*, 1997). Phases of cubic zirconia doped with 23–52 wt.% Pu and zirconolite doped with approximately 6 wt.% Pu were identified in a glass-ceramic wasteform developed for plutonium immobilisation together with calcined HLW from the Idaho Chem-ical Reprocessing Plant (ICPP)(O'Holleran *et al.*, 1997). Synthesis of apatite-based wasteforms for immobilisation of different radionu-clides including actinides has been reviewed (Carpena *et al.*, 1998). Thorium phosphate-diphosphate, $Th_4(PO_4)_4P_2O_7$, TPD, was pro-posed as a durable host phase for tetravalent actinide incorpo-ration by researchers at the Universite de Paris-Sud, Institut de Nucleaire, France (Dacheux *et al.*, 1998a). Different solid solutions $(Th,Pu)_4(PO_4)_4P_2O_7$ and $(Th,Np)_4(PO_4)_4P_2O_7$ were synthesised (Dacheux *et al.*, 1998b; Dacheux *et al.*, 1998c; Dacheux *et al.*, 1999). Formation of $PuPO_4$ with monazite structure instead of the desired $Pu_4(PO_4)_4P_2O_7$ was also observed (Dacheux *et al.*, 1998b).

Synthesis of zirconate pyrochlore $Pu_2Zr_2O_7$ and $Am_2Zr_2O_7$ was achieved in a joint ORNL, USA and JAERI programme (Raison *et al.*, 1999). Polycrystalline zircon pellets doped with 2 and 10 wt.% ^{241}Am were fabricated at the Institute for Transuranium Elements (ITU), Germany (Burghartz *et al.*, 1998). Garnet-structured solid solution, $(Y,Gd)_3(Al,Ga)_5O_{12}$, was suggested for Ga-containing plu-tonium wastes in 1998 (Burakov and Strykanova, 1998).

Synthesis of large scale wasteform blocks immobilising U and Pu in zirconia via self-sustaining reactions was experimentally demonstrated (Ojovan *et al.*, 1999; Kulyako *et al.*, 2001) and synthesis of Synroc using self-propagating high temperature synthesis for waste containing metallic Zr was modelled (Sobolev *et al.*, 1998).

Yttria-stabilised cubic zirconia $(Zr,Y,Np)O_x$ doped with ^{237}Np (10, 20, 30, 40 mol.% NpO_2) was fabricated in Japan (Kinoshita *et al.*, 1998).

Synthetic murataite with the simplified formula $A_4B_2C_7O_{22}$, where "A" = Ca, Mn, TR, U; "B" = Mn, Ti, Zr, U; C = Ti, Al, was proposed as a host phase for actinide immobilisation (Laverov *et al.*, 1998; Stefanovsky *et al.*, 1999).

Ceramics based on Pu-doped (up to 5–6 wt.% ^{239}Pu) gadolinium-aluminum garnet and perovskite were obtained by melting at KRI (Burakov *et al.*, 2000). ^{239}Pu-doped (5–10 wt.% Pu) ceramics such as: pyrochlore, $(Ca,Gd,Pu,U,Hf)_2Ti_2O_7$; zircon, $(Zr,Pu)SiO_4$; cubic zirconia, $(Zr,Gd,Pu)O_2$, were synthesised at KRI in collaboration with LLNL (Burakov and Anderson, 2000).

Good ion radiation damage resistance of gadolinium zirconate, $Gd_2Zr_2O_7$, with pyrochlore structure under Xe ion bombardment was discovered and zirconate pyrochlore was suggested as a host phase for plutonium (Wang *et al.*, 1999). Under ion irradiation gadolinium zirconate changes from pyrochlore to fluorite-type structure but remains crystalline at very high doses.

3.1.3. *Emergence of durability studies*

Studies of chemical durability of cubic zirconia, $(Zr,Y,Np)O_x$ doped with ^{237}Np (20, 30 and 40 mol.% NpO_2) revealed that leaching rates of Np from zirconia were much smaller than those from Synroc (Kinoshita *et al.*, 2000).

MCC-1[1] leach tests at 90°C in deionised water of titanate ceramics based on pyrochlore and zirconolite containing about 15 wt.%

---

[1]MCC-1 test measures the leach durability of wasteforms in static conditions open to the atmosphere and is used to compare wasteforms (Strachan, 2001).

Pu revealed high durability acceptable for Pu immobilisation (Hart *et al.*, 2000).

PNNL researchers investigated ^{238}Pu-doped ceramics based on pyrochlore, $(Ca,Gd,Pu,U,Hf)_2Ti_2O_7$ (11–12 wt.% PuO_2; 19–24 wt.% UO_2), and zirconolite, $Ca(Hf,U,Pu,Gd)Ti_2O_7$ (7.4 wt.% PuO_2; 1.8 wt.% UO_2) (Strachan *et al.*, 2000). Both ceramics were found to be acceptable for reliable immobilisation of weapons plutonium.

^{238}Pu-doped (8.7 wt.% ^{238}Pu) pyrochlore ceramics, $(Ca,Gd, Pu,U,Hf)_2Ti_2O_7$, were synthesised at the Research Institute of Atomic Reactors (RIAR), Russia (Volkov *et al.*, 2001; Lukinykh *et al.*, 2002). ^{239}Pu-doped (5–10 wt.% Pu) ceramics based on zircon, $(Zr,Pu)SiO_4$, and hafnon, $(Hf,Pu)SiO_4$, were obtained at KRI (Burakov *et al.*, 2001a). Loading of 5–6 wt.% Pu into the ceramic produced homogeneous solid solution matrices, whereas 10 wt.% Pu loading caused formation of PuO_2 inclusions.

Zircon, $(Zr,Pu)SiO_4$, doped with approximately 5 wt.% ^{238}Pu and cubic zirconia ceramics, $(Zr,Gd,Pu)O_2$ doped with 10 wt.% ^{238}Pu were made at KRI (Burakov *et al.*, 2001b). Cubic zirconia demonstrated high resistance to radiation damage. The chemical resistance of the zircon-based ceramic, even though affected by radiation damage, remained high.

Radiation damage studies of ^{238}Pu-doped titanate ceramics (Strachan *et al.*, 2002) revealed that damage occurred faster in zirconolite and pyrochlore based ceramics than expected. The overall dissolution rate of damaged pyrochlore in deionised water in a 3-day test at 90°C was 0.04 g/m^2 and for damaged zirconolite was 0.3 g/m^2. Am and Cm solid solutions in cubic zirconia (Raison *et al.*, 2002) revealed different phases in the systems $(Y_{0.4}Zr_{0.6})O_{1.8}$-$(Y_{0.4}Am_{0.6})O_{1.8}$ and $(Y_{0.4}Zr_{0.6})O_{1.8}$-AmO_2 characterised by cubic fluorite-type structure. Phases in the system $CmO_{1.5}$-ZrO_2 were both fluorite-type structure — such as $(Cm_xZr_{1-x})O_{2-2/x}$ and pyrochlore-type, $Cm_2Zr_2O_7$. The californium pyrochlore, $Cf_2Zr_2O_7$, was also synthesised.

Murataite-based ceramics containing 8 and 10 wt.% PuO_2 were obtained in SIA "Radon" using induction cold crucible melting (Stefanovsky *et al.*, 2001). Perovskite, $(Ca,Gd,Pu^{3+})(Ti,Al)O_3$, accompanying murataite, $(Ca,Mn,Gd,Pu)_4(Mn,Ti,Zr,Pu)_2(TiAl,Fe)_7O_{22-x}$, was identified in these ceramics. It was however noted that attempts to obtain murataite via sintering were unsuccessful and so murataite may not be an equilibrium phase. It may be stabilised by melt quenching, trace element(s), or some particular stoichiometry of the composition such that cation ordering is favoured on the basis of size/charge (Hyatt, 2009).

Cathodoluminescence of ^{239}Pu-doped and ^{238}Pu-doped pyrochlore ceramic obtained in RIAR (Zamoryanskaya *et al.*, 2002) revealed formation of uranyl-ion, $(UO_2)^{2+}$, in synthetic pyrochlore, $(Ca,Gd,Pu,U,Hf)_2Ti_2O_7$, as a result of self-irradiation. This observation was correlated with the presence of uranyl-ion in natural metamict uranpyrochlore, $(Ca,Na,U)_2(Ti,Nb,Ta)_2(O,OH,\dots)_7$.

Synroc-type ceramic wasteforms have been synthesised using self-sustaining (self-propagating) high temperature synthesis reactions in a mixture with zirconium-containing radioactive waste (Ojovan *et al.*, 2001). This has experimentally demonstrated the potential use of waste Zr for actinide, and other long-lived radionuclide immobilisation, in a self-sustaining regime.

Different samples of critical-mass-free plutonium ceramics based on solid solutions $(Hf,Pu,Y)O_x$ and $(Hf,Pu)O_2$ were synthesised and studied at the Institute of Inorganic Materials (VNIINM) (Timofeeva *et al.*, 2002).

Zircon single crystals, $(Zr,Pu)SiO_4$, doped with 2.4–2.6 wt.% ^{238}Pu were synthesised at KRI (Hanchar *et al.*, 2003). Radiation damage including cracking, change of colour and cathodoluminescence were observed.

Powder samples of Pu-doped and (Pu + Am)-doped chlorapatite, $Ca_5(PO_4)_3Cl$, and spodiosite, $Ca_2(PO_4)Cl$, (Metcalfe *et al.*, 2003) gave low normalised mass loss for an MCC-1 leach test in demineralised water at 40°C for 28 days: $(3.6–11.9)\cdot10^{-6}$ g/m^2 for Pu and $2.4\cdot10^{-7}$ g/m^2 for Am.

The use of sodium zirconium phosphate (NZP), $NaZr_2(PO_4)_3$, for HLW immobilisation was suggested by Japanese scientists (Seida *et al.*, 2003). The flexibility of the NZP structure to incorporation of different multi-valent chemical elements was noted. Synthesis of different phosphates with kosnarite structure $KU_2(PO_4)_3$,

$RbU_2(PO_4)_3$, $NaPu_2(PO_4)_3$, $KPu_2(PO_4)_3$, $RbPu_2(PO_4)_3$, $NaNp_2$
$(PO_4)_3$, $KNp_2(PO_4)_3$, $RbNp_2(PO_4)_3$, in RIAR, was also reported
(Volkov *et al.*, 2003) and their potential as host phases for actinide
incorporation was experimentally confirmed.

Detailed investigation of the behaviour of ^{238}Pu doped (9.9 wt.%
^{238}Pu) gadolinia-stabilised cubic zirconia, $Zr_{0.79}Gd_{0.14}Pu_{0.07}O_{1.99}$,
revealed it remains crystalline after a cumulative dose $277 \cdot 10^{23}$ alpha-
decays/m^3 (Burakov *et al.*, 2004a).

Pu-doped zirconolite ceramics (containing 10 wt.% $^{239}PuO_2$ or
10 wt.% $^{238}PuO_2$) were obtained by sintering in air at $1,450°C$ to an
average density of more than 93.3% of theoretical (Advocat *et al.*,
2004). PNNL has reported an extensive study of radiation damage effects in ^{238}Pu-doped titanate ceramics based on zirconolite
pyrochlore (Strachan *et al.*, 2004). ^{238}Pu-doped (8.1 wt.% ^{238}Pu)
monazite, $(La,Pu)PO_4$, and separately, $PuPO_4$ containing 7.2 wt.%
^{238}Pu, were synthesised at KRI (Burakov *et al.*, 2004b). La-monazite
under ambient conditions remained crystalline to a cumulative dose
of $119 \cdot 10^{23}$ alpha-decays/m^3, whereas $PuPO_4$ become almost completely amorphous at a dose of only $42 \cdot 10^{23}$ alpha-decays/m^3.

Powder samples based on actinide doped (3.8% plutonium (3+)
chloride and 0.2% americium chloride in simulated waste) chlorapatite, $Ca_5(PO_4)_3Cl$, and spodiosite, $Ca_2(PO_4)Cl$, were made at
PNNL (Metcalfe *et al.*, 2004). Some samples were doped with ^{238}Pu
instead of ^{239}Pu to study radiation damage. No radiation damage was
detected after 532 days (at a dose of $6 \cdot 10^{17}$ alpha-decays/g). SEM
and microprobe analysis of ^{238}Pu-doped pyrochlore ceramics revealed
destruction of the pyrochlore solid solution, $(Ca,Gd,Pu,U,Hf)_2Ti_2O_7$,
and formation of new phases under self-irradiation (Zamoryanskaya
and Burakov, 2004). Single crystals of zircon, $(Zr,Pu)SiO_4$, doped
with 8–14 wt.% ^{239}Pu were obtained and studied in KRI (Hanchar
et al., 2004).

Radiation effects studies in pyrochlore $Am_2Zr_2O_7$ and $Cf_2Zr_2O_7$
revealed conversion of pyrochlore to a fluorite-type structure under
self-irradiation (Sikora *et al.*, 2005). This observation was interpreted
as a consequence of the changing valence state of Cf and Am from
(3+) to (4+).

Radiation effects in ^{238}Pu-doped titanate pyrochlore ceramic have been summarised (Strachan *et al.*, 2005). Despite radiation damage to the pyrochlore structure under self-irradiation this material remained chemically and physically viable for immobilising surplus weapons-grade plutonium.

Recovery of metamict natural zircon in an acidic solution at 175°C was compared with polycrystalline artificial ^{238}Pu-doped zircon which accumulated the same dose under self-irradiation (Geisler *et al.*, 2005). The behaviour of artificial and natural zircon was significantly different. High rate recrystallisation was only observed in ^{238}Pu-doped zircon.

High mechanical resistance of cubic zirconia, $(Zr,Y,Np)O_x$ doped with ^{237}Np (20, 30, 40 mol.% NpO_2) was confirmed by Kinoshita *et al.* (2006).

20-year-old ^{238}Pu-doped Synroc ceramic pellets were characterised recently (Hambley *et al.*, 2008) revealing insignificant crack formation and retained mechanical durability. Nuclear magnetic resonance (NMR) of fully amorphous polycrystalline zircon doped with ^{238}Pu was done by Farnan *et al.* (2007) revealing the effectiveness of NMR for observing radiation damage effects in crystalline materials. Burakov *et al.* (2008) reviewed 5 years of research into radiation damage effects in ceramics and single crystals doped with ^{238}Pu (cubic zirconia, zircon, monazite, Ti-pyrochlore) concluding that cubic zirconia demonstrated extremely high resistance to self-irradiation. Also, it was observed that amorphisation of zircon was not accompanied by destruction of solid solution in contrast to Ti-pyrochlore, which before complete amorphisation demonstrated destruction of solid solution and formation of some new phases. Resistance of monazite-structured phases correlated strongly with the level of actinide loading. Single crystals of zircon and monazite under self-irradiation lost mechanical integrity accompanied with formation of tiny dispersed particles.

The durability of cubic zirconia under high pressure and temperature in a granitic system has been experimentally confirmed (Gibb *et al.*, 2008a; Gibb *et al.*, 2008b). This suggests its potential as a host, e.g., for Pu in very deep disposal in boreholes.

Initial results on development of durable self-glowing crystals doped with small admixtures of plutonium and other actinides (Burakov *et al.*, 2007) revealed their potential as safe, durable actinide-doped crystalline materials.

Ceramics based on ferrite garnet doped with 2 wt.% ^{244}Cm, $Ca_{1.5}Gd_{0.908}Cm_{0.092}Th_{0.5}ZrFe_4O_{12}$, (Lukinykh *et al.*, 2008) became amorphous after a cumulative dose of $76 \cdot 10^{23}$ alpha-decays/m^3.

A review of radiation effects in ^{238}Pu-doped zirconolites (Strachan *et al.*, 2008) reveals that even after radiation-induced swelling saturated, the ceramics remained physically intact with no evidence of microcracking. Data on actinide ceramic wasteforms were summarised in Laverov *et al.* (2008). Intensively self-glowing zircon and xenotyme crystals doped with ^{238}Pu (≤ 0.1 wt.% ^{238}Pu) were successfully synthesised (Burakov *et al.*, 2009). The possibility of obtaining chemically-durable self-glowing materials with relatively low actinide content was demonstrated. Cm-doped ceramics based on pyrochlore of target composition $Gd_{1.935}Cm_{0.065}TiZrO_7$ (Yudintsev *et al.*, 2009) were rendered amorphous at doses of $4.6 \cdot 10^{18}$ alpha-decays/g (0.60 dpa).

3.2. Titanate-based Ceramics

3.2.1. *Synroc*

So called "Synroc" or "synthetic rock" is the well-known titanite polyphase ceramic developed by Australian scientists as an alternative wasteform to glass for the immobilisation of bulk HLW. Detailed descriptions of Synroc compositions and their principle features are summarised by Ringwood *et al.* (1988). Synroc consists of an assemblage of four main titanate minerals: zirconolite, $CaZrTi_2O_7$, Ba-hollandite, $BaAl_2Ti_6O_{16}$; perovskite, $CaTiO_3$ and titanium oxides, TiO_x. Only zirconolite and perovskite can accommodate actinides. The exact proportions of the main phases vary depending on the HLW composition. For example, Synroc-C is designed to contain about 20 wt.% of calcined HLW and it consists of approximately (in wt.%): 30 — hollandite; 30 — zirconolite; 20 — perovskite and 20 —

84   *Crystalline Materials for Actinide Immobilisation*

Ti-oxides and other phases. Immobilising weapons-grade plutonium or transuranium wastes instead of bulk HLW may essentially change the Synroc phase composition to primarily zirconolite-based or a pyrochlore-based ceramic. The starting precursor for Synroc-C fabrication contains ~57 wt.% TiO_2 and 2 wt.% metallic Ti. The metallic titanium provides reducing conditions during ceramic synthesis and helps decrease volatilisation of radioactive cesium. The use of metallic admixtures is not needed if the Synroc precursor is designed solely for actinide immobilisation. Synroc ceramics can be obtained by different methods including hot pressing, cold pressing followed by sintering in air, melting and self-sustaining (self-propagating) high temperature reactions (see Chap. 4). Synroc is a unique ceramic wasteform, which has been tested at laboratory scale in many countries. Samples of Synroc or polycrystalline zirconolite including ones doped with ^{239}Pu, ^{238}Pu, ^{244}Cm have been examined by teams in Australia, US, UK, France, Germany, Japan and Russia (see Sec. 3.1). However, while inactive pilot plant scale trials have been performed no full scale active work has been.

3.2.2. *Ti-pyrochlore*

Titanate pyrochlore as a possible actinide host phase in a tailored ceramic was mentioned as early as 1982, however it did not attract real attention until 1995–1998 (see Sec. 3.1). The initial LLNL and ANSTO idea was to develop zirconolite-based ceramics (similar to Synroc) for immobilisation of excess weapons-grade plutonium and, possibly, ^{235}U. Two neutron absorbers with different chemical behaviour: Gd and Hf were included. If alteration and plutonium release from the ceramic matrix should occur under the unpredictable conditions of a geological repository the Gd^{3+} would follow Pu^{3+}, and Hf^{4+} accompany migration of Pu^{4+} so avoiding any criticality problems. The proposed ceramic precursor simplified basic composition was (in wt.%): UO_2 — 23.7; PuO_2 — 11.9; Gd_2O_3 — 8.0; HfO_2 — 10.7; CaO — 10.0; TiO_2 — 35.7. Experiments on synthesis have demonstrated formation of pyrochlore, $(Ca,Gd,Pu,U,Hf)_2Ti_2O_7$, instead of zirconolite (Ebbinghaus *et al.*,

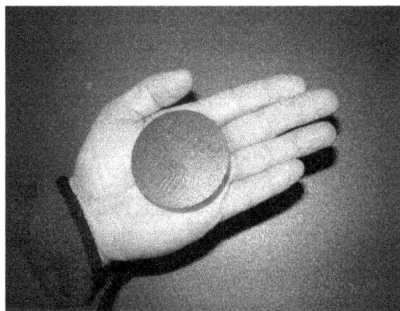

Fig. 3.2.1. Full-size sample of Ce-doped pyrochlore ceramic fabricated at Lawrence Livermore National Laboratory, USA.

1998). A processing route based on cold pressing the oxide precursor followed by sintering in air at 1,350–1,400°C has been developed. Pilot-scale equipment for fabrication of full-size ceramics including Pu-doped samples was developed and fabricated at LLNL. Inactive materials doped with Ce (Fig. 3.2.1) as well as Pu-doped samples were tested.

Under a framework agreement between LLNL and Russian research institutes between 1999–2002 (Burakov and Anderson, 2000; Burakov and Anderson, 2002) samples of Pu-doped pyrochlore

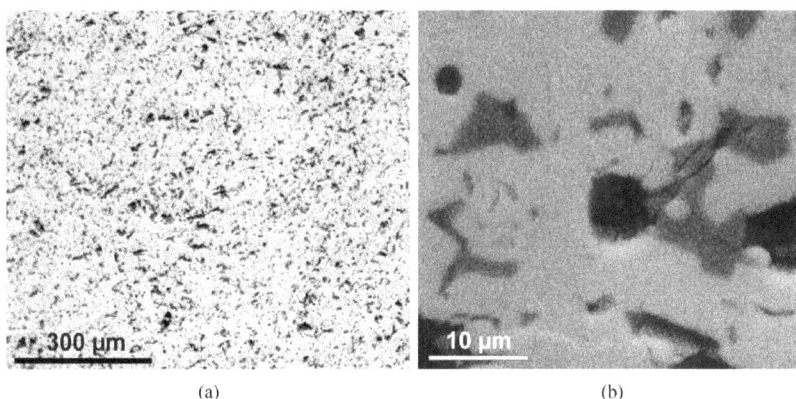

(a) (b)

Fig. 3.2.2. Backscattered electron SEM images of Pu-doped (approximately 10 wt.% ^{239}Pu) pyrochlore ceramic fabricated by sintering in air at 1400°C in V.G. Khlopin Radium Institute, Russia. Light phase is pyrochlore, $(Ca,Gd,Pu,U,Hf)_2Ti_2O_7$; grey phase is rutile, TiO_2, and black areas are pores.

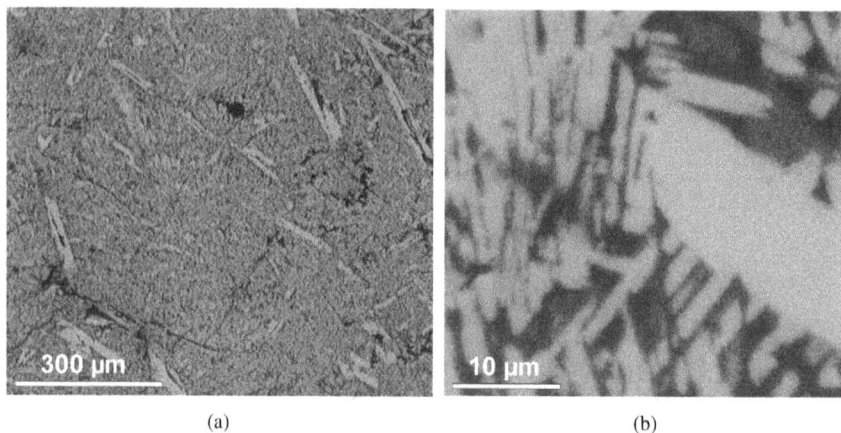

(a)   (b)

Fig. 3.2.3.   Backscattered electron SEM images of Pu-doped (approximately 10 wt.% ^{239}Pu) pyrochlore ceramic fabricated by melting in CO atmosphere at 1380°C at V.G. Khlopin Radium Institute, Russia. Light phase is pyrochlore, $(Ca,Gd,Pu,U,Hf)_2Ti_2O_7$; grey phase was not identified.

ceramic were synthesised at the KRI by sintering in air (Fig. 3.2.2). It was also shown that pyrochlore ceramics could be synthesised by melting at 1,350–1,380°C in an atmosphere of carbon monoxide (CO) (Fig. 3.2.3).

Ceramics based on Ti-pyrochlore are attractive for industrial-scale application because using simple oxide precursors avoids complicated chemical preparation from aqueous actinide-bearing solutions. Dense ceramics are obtained by sintering in air without resorting to hot pressing. However, there are some disadvantages:

(a) The choice of Ti-pyrochlores as appropriate actinide hosts is not justified by study of natural analogues. Ion-exchange adversely affects the weathering of natural pyrochlores (see Sec. 1.4.1);

(b) Enriched uranium is not currently considered a material for immobilisation and it should not be included in pyrochlore ceramic formulations;

(c) Resistance of Ti-pyrochlore to self-irradiation from actinides incorporated in it, is lower than that of zirconate pyrochlores (see Chap. 6).

3.3. Phosphate-based Ceramics

3.3.1. *Monazite*

The principle features of monazite ceramics have been summarised by Boatner and Sales (1988). Use of monazite, $(La,Ce,Gd,...)PO_4$, as a potential actinide host is supported by study of natural analogues (see Sec. 1.4). Partially crystalline monazite-structured solid solutions of lanthanide-actinide phosphates are easily precipitated from liquid wastes by adding phosphoric acid or ammonia-phosphate (see Chap. 4). Further ceramic synthesis from co-precipitated precursors is not accompanied by problematic actinide segregation. Another important advantage is the ability to fabricate dense ceramics by sintering in air at relatively low temperature $(1,200-1,250°C)$. The monazite structure can incorporate high actinide loadings and theoretically it may reveal a continuous solid solution for trivalent actinides. However, synthesis of ^{238}Pu-doped monazite (Burakov *et al.*, 2004b) demonstrated the need to verify the optimal level of actinide loading into a monazite ceramic. Exceeding the critical actinide content may not affect the ceramic behaviour over short periods of time but later it may cause rapid amorphisation, matrix swelling and loss of chemical and mechanical durability. The mechanical durability of monazite ceramic is not high and is similar to that of glass.

3.3.2. *Th-phosphate/diphosphate (TPD)*

Ceramics based on thorium phosphate diphosphate (TPD) with the general formula $Th_{4-x}M_x(PO_4)_4P_2O_7$, where M = U, Np, Pu, were developed by French scientists for the immobilisation of weapons plutonium and other actinides (Dacheux *et al.*, 1998a). TPD can accommodate actinides in thorium sites (in wt.%): U — 47.6; Np — 33.2; Pu — 26.1. Different TPD samples based on single-phase solid solutions $(Th,Pu)_4(PO_4)_4P_2O_7$ and $(Th,Np)_4(PO_4)_4P_2O_7$ have been synthesised (Dacheux *et al.*, 1998b; Dacheux *et al.*, 1998c; Dacheux *et al.*, 1999).

An important advantage of TPD is the simplicity of precursor preparation. Raw TPD precipitate is obtained by mixing concentrated Th-nitrate and phosphoric acid. The ceramic can then be obtained by sintering in air at 1,100–1,350°C.

The mechanical durability of TPD ceramics is similar to that of other phosphate ceramics and comparable or less than that of borosilicate glass.

3.3.3. *Kosnarite and NZP*

Limited studies of ceramics based on kosnarite-structured phosphates have been performed. Sodium zirconium phosphate (NZP), $NaZr_2(PO_4)_3$, not only accommodates actinides substituting for Zr, but also many other radionuclides from HLW (Seida *et al.*, 2003). Such a ceramic wasteform would be attractive for immobilisation of actinide-containing wastes of complex chemical composition. Synthesis of $KU_2(PO_4)_3$, $RbU_2(PO_4)_3$, $NaPu_2(PO_4)_3$, $KPu_2(PO_4)_3$, $RbPu_2(PO_4)_3$, $NaNp_2(PO_4)_3$, $KNp_2(PO_4)_3$, $RbNp_2(PO_4)_3$, was reported (Volkov *et al.*, 2003) although the principal features of ceramics based on actinide-doped NZP or other kosnarite-type host phases require further investigation. These ceramics can be synthesised at low temperature (900–1,000°C) but fabrication of dense NZP ceramics by sintering in air is difficult. Also, the mechanical durability of kosnarite-type ceramic is low — less than that of borosilicate glass.

3.3.4. *Apatite*

Ceramics based on phosphate apatite $Ca_5(PO_4)_3(OH,Cl,F)$ have few advantages over other phosphate ceramics. Their mechanical durability is similar to monazite ceramics and glasses. No data on optimal actinide loading capacity of phosphate apatites are available. There is no doubt however, that it is less than for monazite. Rare-earth silicate apatite (up to phosphor-free composition), $Ca_{4-x}REE_{6+x}(SiO_4)_{6-y}(PO_4)_y(F,OH,O)_2$, has the highest actinide loading capacity and has been suggested for immobilisation of plutonium and other actinides (Ewing *et al.*, 1996; Carpena *et al.*, 1998).

Pure fluorapatite, $Ca_5(PO_4)_3F$, is chemically more durable than chlorapatite, $Ca_5(PO_4)_3Cl$, and, in particular, hydroxyapatite, $Ca_5(PO_4)_3OH$, but all species are relatively easy to dissolve in HCl and other acids.

Similar to monazite partly-crystalline fluorapatite can be obtained by precipitation from solution (see Chap. 4). Dense single phase fluorapatite ceramics are obtained by sintering at relatively low temperature ($1,250$–$1,300°C$). Synthesis of $Ca_9Nd(PO_4)_5SiO_4F_2$ by sintering in air at $1,250°C$ is accompanied by formation of 20–40% minor phase monazite, $NdPO_4$.

Single crystals of fluorapatite used as laser materials are grown from melts suggesting the possibility of using melting for fabrication of fluorapatite-based wasteforms. Successful synthesis of silicate apatite such as $Ca_9Nd(PO_4)_5SiO_4F_2$ by melting a mixture of CaF_2, P_2O_5, $CaCO_3$, Nd_2O_3 and SiO_2 at $1,700°C$ has been reported (Carpena *et al.*, 1998).

Chlorapatite ceramic is attractive for immobilisation of Pu-Am-bearing wastes containing chloride salts. Synthesis of powder samples of chlorapatite doped with Pu and Am at low temperature ($750°C$) in flowing argon (Metcalfe *et al.*, 2003 and 2004) has been demonstrated although dense actinide-doped chlorapatite ceramic was not obtained.

3.4. Ceramics Based on Zirconium and Hafnium Minerals

3.4.1. *Zircon/zirconia and hafnon/hafnia*

Zircon, $ZrSiO_4$, is a well-known refractory material with hardness higher than quartz. Hafnon, $HfSiO_4$, is a crystallochemical analogue of zircon and together they form uninterrupted $HfSiO_4$-$ZrSiO_4$ solid solutions. In contrast with Zr, which is used in metal alloys for nuclear fuel cladding, Hf is a neutron absorber. Therefore, the use of hafnon or zircon/hafnon solid solution ceramics for plutonium immobilisation has the advantage that it avoids criticality issues.

Non-radioactive zircon ceramic is fabricated primarily by grinding natural zircon extracted from placer deposits. Synthesis of zircon,

$ZrSiO_4$, or hafnon, $HfSiO_4$, from oxide precursors ($ZrO_2 + SiO_2$ or $HfO_2 + SiO_2$) is difficult because of the low chemical reactivity of zirconia and hafnia. Fabrication of actinide-doped zircon or hafnon ceramic from oxide precursors is unlikely to be viable because it is almost impossible to avoid unreacted actinide oxide phases in the matrix of the final ceramic.

Use of sol-gel or co-precipitated precursors for synthesis of actinide-doped zircon ceramics is more feasible (see Chap. 4). Solidified actinide-doped gel for zircon synthesis is characterised by unusually high chemical durability (Burakov *et al.*, 2006). A natural analogue has demonstrated geochemical stability for millions of years (see Sec. 1.4.14). Solid Zr-silicate gels from gelation of actinide-bearing solutions might be suitable durable materials for intermediate stabilisation and temporary storage of liquid actinide wastes.

Single-phase zircon ceramic is not attractive for industrial scale actinide immobilisation due to the difficulty of retaining a precise zircon stoichiometry in the starting precursor. It is more reasonable to develop ceramics based on dual-phase composition: zircon and zirconia. A small Zr excess in the precursor simplifies ceramic fabrication and enables full actinide incorporation into the zircon and zirconia host phases avoiding precipitation of separate actinide phases (oxides and silicates). Admixture of small amounts (about 10 wt.%) of monoclinic ZrO_2 increases the resistance of zircon ceramic to thermal shock and improves its mechanical durability (Garvie *et al.*, 1989).

Although single-crystal zircon doped with up to 14 wt.% Pu has been successfully synthesised (Hanchar *et al.*, 2004), synthesis of homogeneous single-phase zircon and hafnon ceramics doped with 10 wt.% Pu (Burakov *et al.*, 2001a) has not been achieved. The ceramics contained PuO_2 inclusions (Fig. 3.4.1). Plutonia inclusions in zircon and hafnon ceramics were not present at lower doping levels (5–6 wt.% Pu) or after change of ceramic formulation from single-phase to biphasic zircon/zirconia or hafnon/hafnia compositions (Fig. 3.4.2).

Formation of zircon in a sol-gel precursor usually starts at 1,300°C (see Chap. 4), however the highest zircon yield is achieved at 1,500–1,600°C. Synthesis of Ce-doped zircon from sol-gel precursors sintered at 1,400 to 1,600°C (at atmospheric pressure) decreased Ce

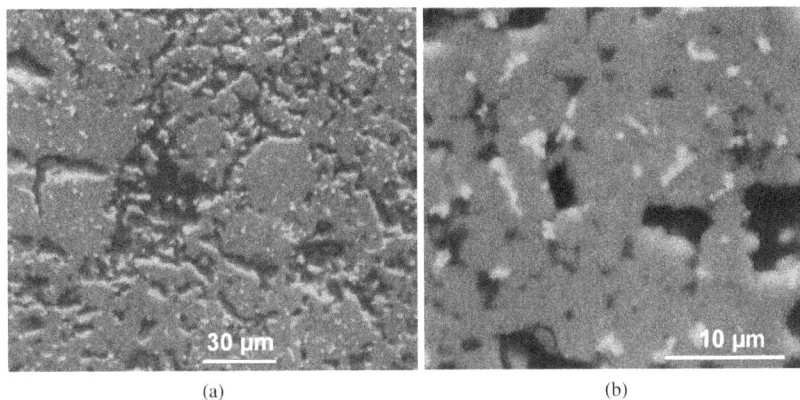

(a)   (b)

Fig. 3.4.1. Backscattered electron SEM images of Pu-doped ceramic based on zircon, $(Zr,Pu)SiO_4$, obtained in V.G. Khlopin Radium Institute (Burakov *et al.*, 2001a). Poor ceramic synthesis led to the presence of PuO_2 inclusions (white phase). Zircon (grey matrix) incorporated only approximately 7 wt.% Pu. Black areas are pores.

incorporation into the zircon structure from 1,550 to 1,600°C. The optimal temperature for synthesis of actinide-doped zircon is thus less than 1,500°C.

Pure zircon melts incongruently or dissociates at 1,670–1,700°C (Pena and De Aza, 1984; Kanno, 1989), forming silica melt with inclusions of crystalline monoclinic or tetragonal zirconia. Different admixtures incorporated into zircon may lower the temperature of zircon dissociation (Pena and De Aza, 1984). Thermal destruction of zircon doped with Ce, Gd and, possibly, actinides does not cause formation of separate phases of these doping elements. Solidified silica melt is free of dopants and on cooling all admixtures are incorporated into the crystalline structure of zirconia (Fig. 3.4.3). The structure of melted zircon is similar to that of a sintered sol-gel precursor for zircon synthesis — before formation of the zircon (Fig. 4.1.5).

Zircon/zirconia ceramics doped with ^{238}Pu reveal high chemical durability under self-irradiation (Geisler *et al.*, 2005; Burakov *et al.*, 2008). Even fully amorphous ^{238}Pu-doped samples were characterised by the absence of visible cracks and low leach rate (see Chap. 5).

Monocrystalline zircon also has potential as a durable self-glowing material (see Chap. 2).

92   *Crystalline Materials for Actinide Immobilisation*

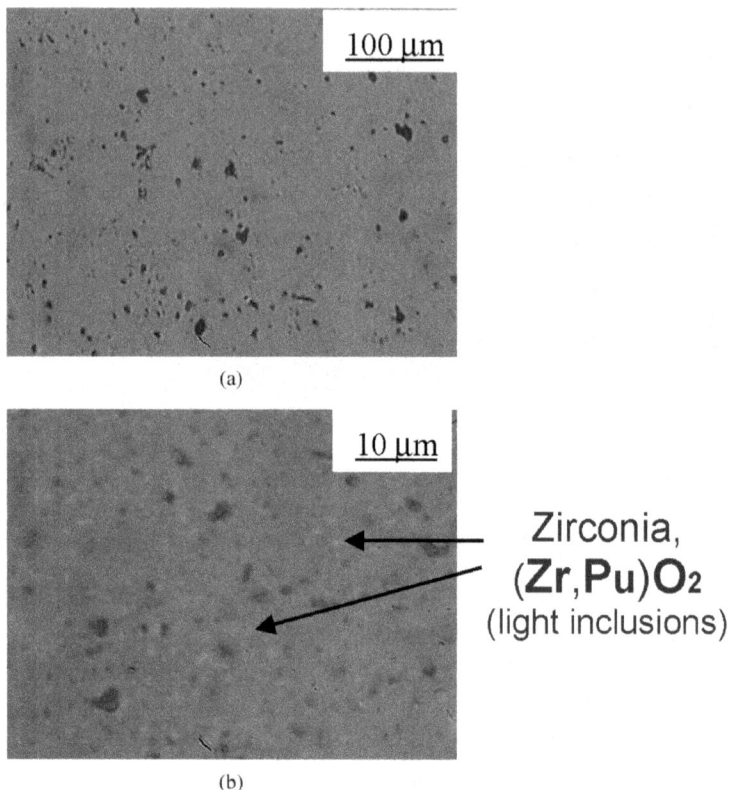

(a)

Zirconia,
(Zr,Pu)O₂
(light inclusions)

(b)

Fig. 3.4.2.   Reflected light images of ^{238}Pu-doped zircon/zirconia ceramic. ^{238}Pu content in the zircon is approximately 4.7 wt.% (Burakov *et al.*, 2001b). Black areas are pores.

3.4.2. *Cubic zirconia (tazheranite) and hafnia*

Polycrystalline cubic zirconia stabilised by yttrium, calcium or rare-earths is a well-known structural and refractory ceramic (melting between 2,700–2,900°C). The high mechanical durability of cubic zirconia ceramics (whose hardness is greater than quartz) allows their use for mortars and pestles and milling media for ball mills. Cubic hafnia is the crystallochemical analogue of cubic zirconia with which it forms uninterrupted hafnia-zirconia solid solutions. Since Hf is a neutron absorber the use of hafnia ceramics (or ceramics based on zirconia/hafnia solid solutions) for plutonium immobilisation avoids

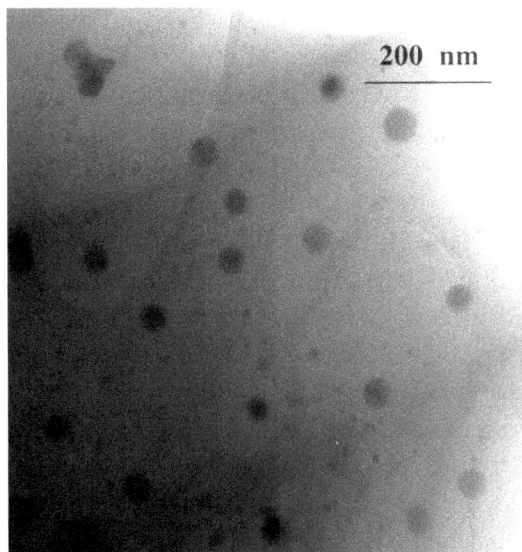

Fig. 3.4.3.   TEM image of Gd-doped zircon after incongruent melting at 1720°C in air. Rounded inclusions are crystalline Gd-doped tetragonal zirconia, $(Zr,Gd)O_2$. Melted matrix is Gd-free amorphous silica.

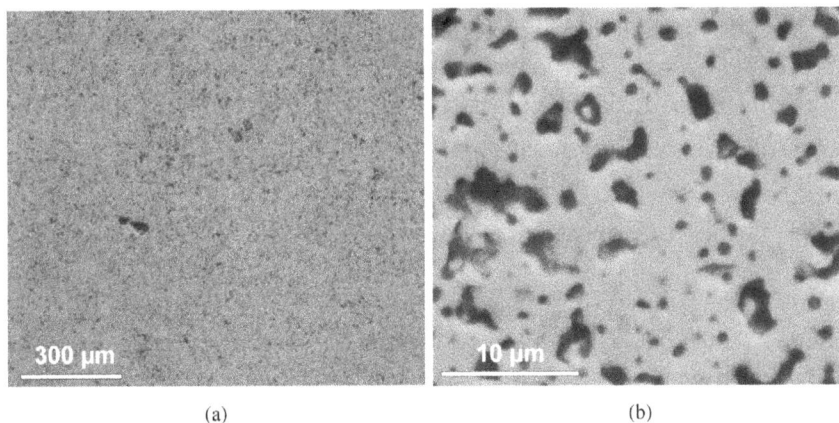

(a)   (b)

Fig. 3.4.4.   Backscattered electron SEM images of ceramic based on gadolinia-stabilised cubic zirconia, $(Zr,Gd,Pu)O_2$, obtained at V.G. Khlopin Radium Institute by sintering in air at 1500°C (Burakov and Anderson, 2002). ^{239}Pu content is 10.3 wt. % and Gd — 20.9 wt.%. Black areas are pores.

any criticality problems. An important advantage of cubic zirconia is its high capacity to incorporate tetra- and trivalent actinides. Yttria-stabilised cubic zirconias, $(Zr,Y,Np)O_2$, doped with ^{237}Np from 10 to 40 mol.% NpO_2 have been fabricated (Kinoshita *et al.*, 1998) demonstrating the feasibility of neptunium immobilisation. Samples of gadolinia-stabilised cubic zirconia, $(Zr,Gd,Pu)O_2$, doped with approximately 10 wt.% ^{239}Pu (Fig. 3.4.4); 10 wt.% ^{238}Pu (Burakov *et al.*, 2001; Burakov *et al.*, 2004a) and yttria-stabilised, $(Zr,Y,Am)O_2$, doped with 20 wt.% ^{243}Am have been fabricated (Anderson and Burakov, 2004). Cubic (fluorite-type) $(Y_{0.4}Zr_{0.6})O_{1.8}$-$(Y_{0.4}Am_{0.6})O_{1.8}$ and $(Y_{0.4}Zr_{0.6})O_{1.8}$-AmO_2 solid solutions have also been made (Raison *et al.*, 2002).

The high resistance of cubic zirconia to self-irradiation (see Chap. 5) and its refractoriness make it potentially suitable for advanced plutonium fuel as well as for transmutation targets (see Chap. 2). Zirconia-hafnia ceramics can be obtained by different methods including hot pressing, cold pressing followed by sintering in air and self-sustaining (self-propagating) high temperature reactions (see Chap. 4). ZrO_2 ceramics typically require sintering temperatures $>1,500°C$ to achieve high density.

Monocrystalline cubic zirconia obtained by cold crucible melting, CCM (see Chap. 4) is a popular artificial gemstone as well as having potential as a durable luminescence and self-glowing material (see Chap. 2).

3.5. Garnet/Perovskite

Some species of synthetic monocrystalline garnet such as $Y_3Al_5O_{12}$ (YAG) and $Gd_3Ga_5O_{12}$ (GGG) are well-known artificial gemstones with hardness higher than quartz and without cleavage problems. However, these materials were developed originally for use in the microelectronics industry and as laser compounds (for example, Nd-doped YAG). Also, monocrystalline aluminate garnet may be a suitable durable luminescence and self-glowing material (see Chap. 2). The garnet structure $A_3B_2(XO_4)_3$ has a high capacity to incorporate various chemical elements in three types of structural cation

positions A, B and X, which are responsible for accommodation of
different chemical elements:

$A = Na^+, Ca^{2+}, Sr^{2+}, Ba^{2+}, Cd^{2+}, Fe^{2+}, Mg^{2+}, Y^{3+}, Ln^{3+}, Zr^{4+}, Hf^{4+};$

$B = Fe^{2+}, Mn^{2+}, Mg^{2+}, Al^{3+}, Ga^{3+}, Y^{3+}, Sc^{3+}, Fe^{3+}, Cr^{3+}, Ln^{3+},$
$\quad Si^{4+}, Ge^{4+}, Sn^{4+}, Zr^{4+}, Hf^{4+};$

$X = Al^{3+}, Fe^{3+}, Ga^{3+}, Si^{4+}, Ge^{4+}, Ti^{4+}, Sn^{4+}.$

The same elements can occupy different positions without affecting the stability of the garnet crystalline structure. High quality
monocrystals of garnet are produced by melting methods at temperatures up to 1,950°C (the melting point of undoped YAG). The
change of composition associated with actinide doping may decrease
its melting point. Polycrystalline YAG can be obtained by solid-state
reaction at 900°C (Bondar *et al.*, 1984). Spherical particles (from 100
nm to 2 micron in size) of rare-earth gallium garnet have been synthesised by a glycothermal method at 300°C (Inoue *et al.*, 1998).

Although the garnet structure is able to incorporate many multi-
valent elements the synthesis of single-phase garnet-based ceramic
wasteforms is not trivial. For example, formation of garnet phase in
melted Ce-doped precursor of chemical composition calculated for
100% garnet stoichiometry $(Y,Gd,Ce)_3(Al,Ga)_5O_{12}$ was accompanied by crystallisation of perovskite, $(Y,Gd,Ce)AlO_3$ (Fig. 3.5.1(a)).
At the same time synthesis of perovskite ceramic by melting Pu-
doped precursor of chemical composition calculated for 100% perovskite stoichiometry, $(Gd,Ce,Ca)(Al,Ga,Pu,Sn)O_3$, formed biphasic
perovskite/garnet ceramics (Fig. 3.5.1(b)). Other ceramics designed
to have single-phase garnet compositions $(Ca,Gd,Pu)Zr_2Fe_3O_{12}$ and
$Na_2(Gd,Ce,Zr,Pu)(Ga,Al)_5O_{12}$ and obtained by sintering and melting respectively were also biphasic (Figs. 3.5.2. and 3.5.3).

Incorporation of tetravalent actinides into garnet ceramics
requires the use of additions providing charge compensation in the
garnet structure. Sn in U-doped gadolinium-gallium garnet (GGG)
may increase uranium incorporation into the garnet structure from
less than 0.05 wt.% to 3–4 wt.% (Burakov *et al.*, 1999).

Garnet ceramics are attractive for immobilisation of actinide-
bearing wastes of complex chemical compositions in the presence of

96   *Crystalline Materials for Actinide Immobilisation*

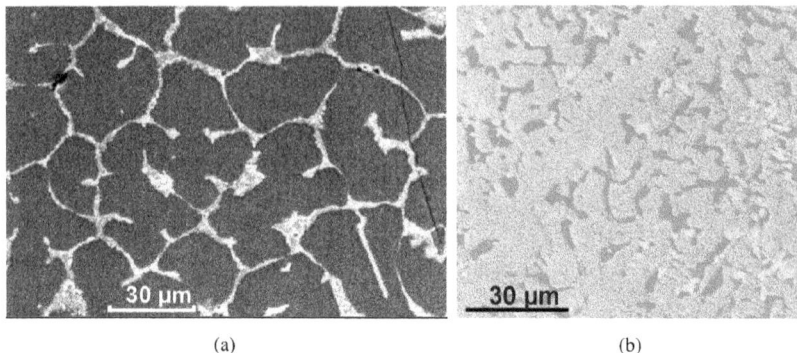

(a)   (b)

Fig. 3.5.1.   Backscattered electron SEM images of aluminate garnet/perovskite ceramics obtained by melting in air: (a) — from Ce-doped precursor with garnet stoichiometry (Burakov and Strykanova, 1998) and (b) — from Pu-doped precursor with perovskite stoichiometry (Burakov *et al.*, 2000). Ce-doped garnet, $(Y,Gd,Ce)_3(Al,Ga)_5O_{12}$ (dark matrix in (a)) contains 0.1–2.0 wt.% Ce, but perovskite, $(Y,Gd,Ce)AlO_3$ (light phase) contains 11–26 wt.% Ce. Pu-doped garnet, $(Gd,Ce,Ca,Sn,Pu)_3(Al,Ga)O_{12}$ (dark phase in (b)) contains 5.3 wt.% Pu, but perovskite, $(Gd,Ce,Ca)(Al,Ga,Pu,Sn)O_3$ (light matrix) — 6.5 wt.% Pu.

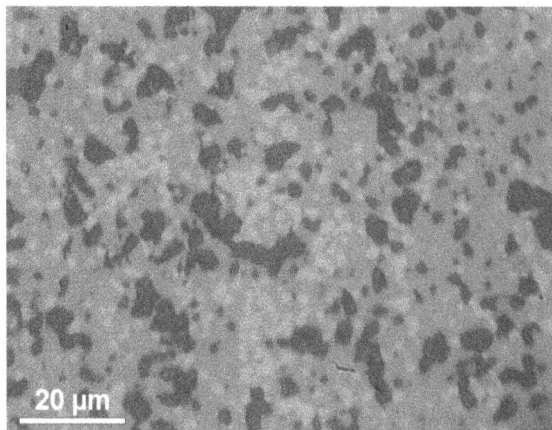

Fig. 3.5.2.   Reflected optical microscope image of ceramic based on garnet, $(Ca,Gd,Pu)Zr_2Fe_3O_{12}$, doped with 4–7 wt.% Pu. Light-grey inclusions are cubic $(Pu,Gd,Zr)O_2$ with fluorite structure and Pu content 22–24 wt.%. Black areas are pores. Sample was obtained in V.G. Khlopin Radium Institute by sintering in air at 1300°C using the route of Dr. S.V. Yudintsev of Institute of Geology of Ore Deposits (IGEM), Russia.

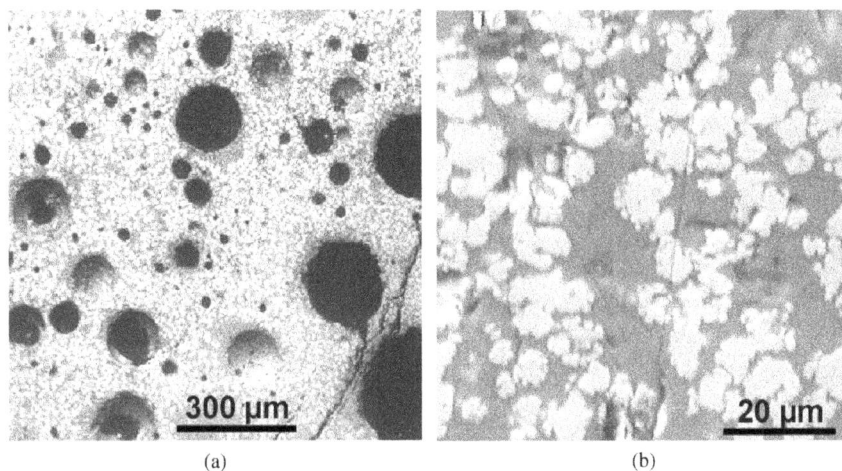

Fig. 3.5.3.   Backscattered electron SEM images of Pu-doped ceramic based on garnet, $Na_2(Gd,Ce,Zr,Pu)(Ga,Al)_5O_{12}$ (grey matrix), and zirconia, $(Zr,Gd,Pu)O_2$ (light inclusions), obtained by melting (Burakov and Anderson, 2002). Garnet phase contains 0.7 wt.% [239]Pu and 6.6 wt.% Na. Content of Pu in zirconia is 16.4 wt.%. Black areas are pores.

non-radioactive species such as Al, Ga, Zr, Fe but low actinide content (less 1 wt.%). Sometimes these wastes (in the form of sludges) are inhomogeneous and the flexibility of garnet formulations is beneficial. To avoid formation of separate actinide phases during synthesis the ceramic formulation uses excess of the main inactive elements in the particular waste stream in the garnet structure. For example, waste streams enriched by aluminium salts can be converted into ceramics based on aluminate garnet and perovskite. Any variation in starting precursor concerning garnet stoichiometry will be compensated by formation of perovskite phase. Waste sludges containing precipitates of iron hydroxides and zirconia would be converted into ferrate garnet ceramic with minor zirconia, able to host any actinides not incorporated into the garnet structure.

Garnet ceramics can be fabricated by, e.g., sintering, melting or self-sustaining (self-propagating) high temperature reactions (see Chap. 4). CCM is an attractive route for obtaining large (ten cm) scale samples of garnet ceramics (Figs. 3.5.4 and 3.5.5).

98   *Crystalline Materials for Actinide Immobilisation*

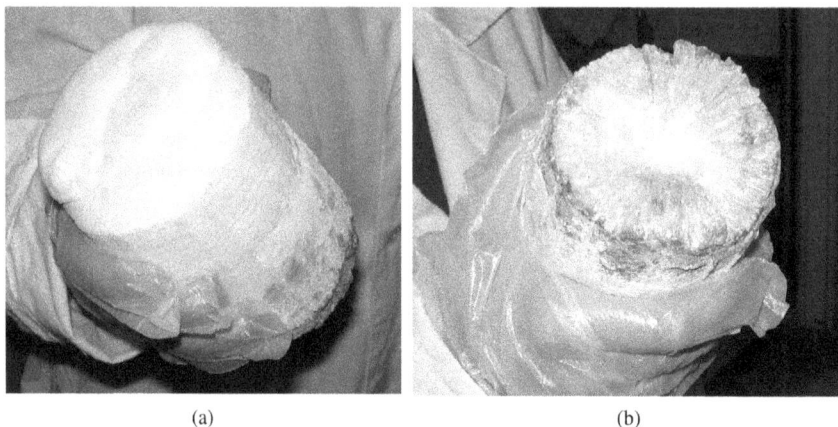

(a)   (b)

Fig. 3.5.4.   Large cylinders (2.5 kg) of polycrystalline $Y_3Al_5O_{12}$ (YAG) obtained by melting in air at $2000°C$ in a CCM. Courtesy of M.V. Zamoryanskaya and B.T. Melekh of the Ioffe Physico-Technical Institute, St. Petersburg, Russia.

Fig. 3.5.5.   Fragment of undoped $Y_3Al_5O_{12}$ (YAG) obtained by melting in air at $2000°C$ in a CCM. Courtesy of M.V. Zamoryanskaya and B.T. Melekh of the Ioffe Physico-Technical Institute, St. Petersburg, Russia.

Table 3.5.1. Summary of principal features of potential ceramics for actinide immobilisation.

Ceramic	Simplified formula	Mechanical durability	Chemical resistance in different media	Immobilisation of actinide-bearing wastes of complex chemical composition	Matrix of fuel or transmutation target
Synroc	$CaZrTi_2O_7$, $BaAl_2Ti_6O_{16}$, $CaTiO_3$, TiO_x	Average	High	Yes	No
Zirconolite	$CaZrTi_2O_7$	Average	High	No	No
Ti-pyrochlore	$Gd_2Ti_2O_7$	Average	Average	Yes	No
Zr-pyrochlore	$Gd_2Zr_2O_7$	Average	Average	No	Unclear
Zircon/zirconia and hafnon/hafnia	$(Zr,Hf)SiO_4/$ $(Zr,Hf)O_2$	High	High	No	No
Cubic zirconia and hafnia	$(Zr,Hf,\ldots)O_2$	High	High	No	Yes
Monazite	$(Ce,La,Gd,Eu,\ldots)PO_4$	Low	High	No	No
Xenotime	YPO_4	Low	Average	No	No
TPD	$Th_4(PO_4)_4P_2O_7$	Low	High	No	No
Kosnarite and NZP	$(K,Na)Zr_2(PO_4)$	Low	Average	Yes	No
Al-garnet	$(Y,Gd,\ldots)_3Al_5O_{12}$	High	High	Yes	Unclear
Al-perovskite	$(Y,Gd,\ldots)AlO_3$	Average	Average	Yes	No
Fe-garnet	$(Y,Gd,\ldots)_3Fe_5O_{12}$	Average	Average	Yes	No
Murataite	$(Ca,Gd,\ldots)_4(Mn,Zr\ldots)_2$ $(TiAl,Fe)_7O_{22}$	Average	Average	Yes	No
F-apatite	$Ca_5(PO_4)_3F$	Low	Average	Unclear	No
Cl-apatite	$Ca_5(PO_4)_3Cl$	Low	Average	Yes	No
Si-apatite	$Ca_{4-x}REE_{6+x}$ $(SiO_4)_{6-y}(PO_4)_y(F,O)_2$	Average	Average	Unclear	No

3.6. Summary

An extensive overview of ceramic wasteforms developed worldwide reveals:

1. Only two ceramics, Synroc and Ti-pyrochlore, have been suffi-ciently tested at pilot scale and are close to being adopted by the nuclear industry. However, no actinide ceramic wasteforms have been used anywhere.
2. The choice of ceramic wasteform is affected by the bulk chemical composition of the actinide-bearing wastes. Some ceramics are not flexible enough to be able to immobilise waste of complex chemical compositions (see Table 3.5.1). Those ceramics which are proposed for accommodation of complex wastes in most cases will have multi-phase compositions (e.g., Synroc; garnet/perovskite; garnet/zirconia).
3. Only cubic zirconia ceramic could be considered as a universal material applicable to actinide wastesforms; inert matrix fuels and targets for actinide transmutation.
4. Only a few ceramics such as zircon/zirconia; hafnon/hafnia; cubic zirconia and hafnia and aluminate garnet are characterised by high mechanical durability (with hardness greater than quartz).

References

Abraham M.M. and Boatner L.A. (1982) Electron-paramagnetic-resonance inves-tigations of ^{243}Cm^{3+} in LuPO$_4$ single crystals, *Phys. Rev., B*, **26**, 1434–1437.
Advocat T., Jorion F., Marcillat T., Leturcq G., Deschanels X., Boubals J.M., Bojat L., Nivet P. and Peuget S. (2004) Fabrication of $^{239/238}$Pu-zirconolite ceramic pellets by natural sintering, *Scientific Basis for Nuclear Waste Man-agement XXVII, Mater. Res. Soc. Symp. Proc.* **807**, 267–272.
Anderson E.B., Burakov B.E. and Vasiliev V.G. (1993) A creation of crystalline matrix for actinide waste in Khlopin Radium Institute. *Proc. Int. Conf. SAFE WASTE'93*, Avignon, France, 13 to 19 June, 1993, **2**, 29–33.
Anderson E.B. and Burakov B.E. (2004) Ceramics for the immobilization of plu-tonium and americium: Current progress of R&D of the V.G. Khlopin Radium Institute, *Scientific Basis for Nuclear Waste Management XXVII, Mater. Res. Society Symp. Proc.* **807**, 207–212.

Boatner L.A. (1978) Letter to the US Department of Energy, Office of Basic
  Energy Sciences, Division of Material Sciences on 28 April detailing possible
  uses of monazite as an alternative to borosilicate glass.
Boatner L.A., Beall G.W., Abraham M.M., Finch C.B., Huray P.G. and Rap-
  paz M. (1980) Monazite and other lanthanide orthophosphates as alternative
  actinide waste forms. in *Scientific Basis for Nuclear Waste Management* ed.
  C.J.M. Northrup Jr. (Plenum Press, New York) **2**, 289–296.
Boatner L.A. and Sales B.C. (1988) in *Radioactive Waste Forms for the Future*,
  eds. W. Lutze and R.C. Ewing (North-Holland Physics Publishing, the
  Netherlands), pp. 495–564.
Bondar I.A., Koroleva L.N. and Bezruk E.T. (1984) Physico-chemical properties
  of yttrium aluminates and gallates, *J. Inorganic Materials* **20**(2), 257–261 (in
  Russian).
Burakov B.E., Britvin S.N., Miheeva E.E., Ilyinskiy G.A., Anderson E.B.,
  Pazukhin E.M., Nikolaeva L.D., Zirlin V.A. and Bogdanova A.N. (1991) Inves-
  tigation of artificial zircon from Chernobyl "lava", *Notes of All-Union Min-
  eralogical Society*, #6, pp. 39–44 (in Russian).
Burakov B. (1993) A study of high-uranium technogeneous zircon $(Zr,U)SiO_4$
  from Chernobyl "lavas" in connection with the problem of creating a
  crystalline matrix for high-level waste disposal, *Proc. Int. Conf. SAFE
  WASTE'93*, Avignon, France, 13 to 18 June 1993, **2**, 19–28.
Burakov B.E., Anderson E.B., Zamoryanskaya M.V., Yagovkina M.A., Strykanova
  E.E. and Nikolaeva E.V. (2001) Synthesis and study of ^{239}Pu-doped ceramics
  based on zircon, $(Zr,Pu)SiO_4$, and hafnon, $(Hf,Pu)SiO_4$, *Scientific Basis for
  Nuclear Waste Management XXIV, Mater. Res. Society Symp. Proc.* **663**,
  307–313.
Burakov B.E. and Strykanova E.E. (1998) Garnet solid solution of $Y_3Al_5O_{12}$-
  $Gd_3Ga_5O_{12}$-$Y_3Ga_5O_{12}$ (YAG-GGG-YGG) as a prospective crystalline host-
  phase for Pu immobilization in the presence of Ga, *Proc. Int. Symp. Waste
  Management'98*, Tucson, AZ, USA, CD.
Burakov B.E., Anderson E.B., Knecht D.A., Zamoryanskaya M.V., Strykanova
  E.E. and Yagovkina M.A. (1999) Synthesis of garnet/perovskite-based
  ceramic for the immobilization of Pu-residue wastes, *Scientific Basis for
  Nuclear Waste Management XXII, Mater. Res. Society Symp. Proc.* **556**,
  55–62.
Burakov B.E., Anderson E.B., Zamoryanskaya M.V. and Petrova M.A. (2000)
  Synthesis and study of ^{239}Pu-doped gadolinium-aluminum garnet, *Scientific
  Basis for Nuclear Waste Management XXIII, Mater. Res. Society Symp. Proc.*
  608, 419–422.
Burakov B.E. and Anderson E.B. (2000) Summary of Pu ceramics developed
  for Pu immobilization (B338247, B501118), in *Excess Weapons Plutonium
  Immobilization in Russia*, eds. L.J. Jardine, G.B. Borisov, Proc. Meet. for
  Coordination and Review of Work, St. Petersburg, Russia, (1999), UCRL-ID-
  138361, pp. 167–179.
Burakov B.E., Anderson E.B., Zamoryanskaya M.V., Nikolaeva E.V., Strykanova
  E.E. and Yagovkina M.A. (2001) Investigation of zircon/zirconia ceramics

102   *Crystalline Materials for Actinide Immobilisation*

doped with ^{239}Pu and ^{238}Pu, *Proc. Intern. Conf. GLOBAL'05*, Paris, France, (2001), CD, paper 006.

Burakov B.E. and Anderson E.B. (2002) Summary of Pu ceramics developed for Pu immobilization (B506216, B512161), in *Review of Excess Weapons Disposition: LLNL Contract Work in Russia*, eds. L.J. Jardine, G.B. Borisov, Proc.3rd Ann. Meet. for Coordination and Review of LLNL Work, St. Petersburg, Russia, (2002), UCRL-ID-149341, pp. 265–270.

Burakov B.E., Yagovkina M.A., Zamoryanskaya M.V., Kitsay A.A., Garbuzov V.M., Anderson E.B. and Pankov A.S. (2004a) Behavior of ^{238}Pu-doped cubic zirconia under self-irradiation, *Scientific Basis for Nuclear Waste Management XXVII*, Mater. Res. Society Symp. Proc. **807**, 213–217.

Burakov B.E., Yagovkina M.A., Garbuzov V.M., Kitsay A.A. and Zirlin V.A. (2004b) Self-irradiation of monazite ceramics: Contrasting behavior of PuPO$_4$ and (La,Pu)PO$_4$ doped with Pu-238, *Scientific Basis for Nuclear Waste Management XXVIII*, Mater. Res. Society Symp. Proc. **824**, 219–224.

Burakov B.E., Smetannikov A.Ph. and Anderson E.B. (2006) Investigation of natural and artificial Zr-silicate gels, *Scientific Basis for Nuclear Waste Management XXIX*, Mater. Res. Soc. Symp. Proc. **932**, 1017–1024.

Burakov B.E., Garbuzov V.M., Kitsay A.A., Zirlin V.A., Petrova M.A., Domracheva Ya.V., Zamoryanskaya M.V., Kolesnikova E.V., Yagovkina M.A. and Orlova M.P. (2007). The use of cathodoluminescence for the development of durable self-glowing crystals based on solid solutions YPO$_4$-EuPO$_4$, *Semiconductors* **41**(4), 427–430.

Burakov B.E., Yagovkina M.A., Zamoryanskaya M.V., Garbuzov V.M., Zirlin V.A. and Kitsay A.A. (2008) Self-irradiation of ceramics and single crystals doped with Pu-238: Summary of 5 years of research of the V.G. Khlopin Radium Institute, *Scientific Basis for Nuclear Waste Management*, Mater. Res. Soc. Symp. Proc. **1107**, 381–388.

Burakov B.E., Domracheva Ya.V., Zamoryanskaya M.V., Petrova M.A., Garbuzov V.M., Kitsay A.A. and Zirlin V.A. (2009) Development and synthesis of durable self-glowing crystals doped with plutonium. *J. Nucl. Mater.* **385**, 134–136.

Burghartz M., Matzke Hj., Leger C., Vambenepe G. and Rome M. (1998) Inert matrices for the transmutation of actinides: Fabrication, thermal properties and radiation stability of ceramic materials, *J. Alloys Compounds* **271–273**, 544–548.

Carpena J., Audubert F., Bernache D., Boyer L., Donazzon B., Lacout J.L. and Senamaud N. (1998) Apatitic waste forms: Process overview, *Mater. Res. Soc. Symp. Proc. Scientific Basis for Nuclear Waste Management XXI*, **506**, 543–549.

Carroll D. (1963) The system PuO$_2$-ZrO$_2$, *J. Am. Ceram. Soc.* **46**, 194–195.

Clinard F.W. Jr., Hobbs L.W., Lands C.C., Peterson D.E., Rohr D.L. and Roof R.B. (1982) Alpha decay self-irradiation damage in ^{238}Pu-substituted zirconolite, *J. Nucl. Mater.* **105**, 248–256.

Caurant D., Loiseau P., Majérus O., Aubin Chevaldonnet V., Bardez I. and Quintas A. (2009) *Glasses, Glass-Ceramics and Ceramics for Immobilization of Highly Radioactive Nuclear Wastes* (Nova, New York).

Dacheux N., Podor R., Chassigneux B., Brandel V. and Genet M. (1998a) Actinides immobilization in new matrices based on solid solutions: $Th_{4-x}M_x(PO_4)_4P_2O_7$, $(M^{IV} = {}^{238}U, {}^{239}Pu)$, *J. Alloys Compounds* **271–273**, 236–239.

Dacheux N., Podor R., Brandel V. and Genet M. (1998b) Investigations of systems $ThO_2-MO_2-P_2O_5$ (M = U, Ce, Zr, Pu). Solid solutions of thorium-uranium (IV) and thorium-plutonium (IV) phosphate-diphosphates, *J. Nucl. Mater.* **252**, 179–186.

Dacheux N., Thomas A.C., Brandel V. and Genet M. (1998c) Investigations of the system $ThO_2-NpO_2-P_2O_5$. Solid solutions of thorium-neptunium (IV) phosphate-diphosphate, *J. Nucl. Mater.* **257**, 108–117.

Dacheux N., Thomas A.C., Chassigneux B., Pichot E., Brandel V. and Genet M. (1999) Study of $Th_4(PO_4)P_2O_7$ and solid solutions with U(IV), Np(IV) and Pu(IV): Synthesis, characterization, sintering and leaching tests, *Scientific Basis for Nuclear Waste Management XXII, Mater. Res. Soc. Symp. Proc.* **556**, 85–92.

Degueldre C., Kasemeyer U., Botta F. and Ledergerber G. (1996) Plutonium incineration in LWR's by a once-through cycle with a rock-like fuel, *Scientific Basis for Nuclear Waste Management XIX, Mater. Res. Soc. Symp. Proc.* **412**, 15–23.

Degueldre C., Heimgartner P., Ledergerber G., Sasajima N., Hojou K., Muromura T., Wang L., Gong W. and Ewing R. (1997) Behaviour of zirconia based fuel material under Xe irradiation, *Mater. Res. Society Symp. Proc.* **439**, 625–632.

Donald I.W., Metcalfe B.L. and Taylor R.N.J. (1997) Review: The immobilization of high level radioactive wastes using ceramics and glasses. *J. Mater. Sci.* **32**, 5851–5887.

Ebbinghaus B., VanKonynenburg R.A., Ryerson F.J., Vance E.R., Stewart M.W.A., Jostsons A., Allender J.S., Rankin T. and Gongdon J. (1998) Ceramic formulation for the immobilization of plutonium, *Proc. Int. Symp. Waste Management'98*, Tucson, AZ, USA, (1998), CD, Rep. 65–04.

Ewing R.C., Lutze W. and Weber W.J. (1995) Zircon: A host-phase for the disposal of weapons plutonium, *J. Mater. Res.* **10**, 243–246.

Ewing R.C., Weber W.J. and Lutze W. (1996) Crystalline ceramics: Waste forms for the disposal of weapons plutonium, in *Disposal of Weapon Plutonium*, eds. E.R. Merz and C.E. Walter, NATO ASI Series (Kluwer Academic Publisher, Dordrecht), pp. 65–83.

Ewing R.C. (1999) Nuclear waste forms for actinides. *Proc. Natl. Acad. Sci.*, USA, **96**, 3432–3439.

Exharos G.J. (1984) Induced swelling in radiation damaged $ZrSiO_4$. *Nucl. Instr. Meth. Phys. Res.* **B1**, 538–541.

Farnan I., Cho H. and Weber W. (2007) Quantification of actinide α-radiation damage in minerals and ceramics, *Nature* **445**, 190–193.

104   *Crystalline Materials for Actinide Immobilisation*

Furuya H., Muraoka S. and Muromura T. (1996) Feasibility of rock-like fuel and
glass waste form for disposal of weapons plutonium, in *Disposal of Weapon
Plutonium*, eds. E.R. Merz and C.E. Walter, NATO ASI Series (Kluwer Academic Publisher, Dordrecht), pp. 107–121.

Garvie R.C., Drennan J., Goss M.F., Marshall S. and Urbani C. (1989) Design
and application of a zircon advanced refractory ceramic, in *Zircon-Science
and Engineering*, ed. S. Somuya, Japan, pp. 299–313.

Geisler T., Burakov B., Yagovkina M., Garbuzov V., Zamoryanskaya M., Zirlin
V. and Nikolaeva L. (2005) Structural recovery of self-irradiated natural and
[238]Pu-doped zircon in an acidic solution at 175°C, *J. Nucl. Mater.* **336**, 22–30.

Gibb F.G.F., Burakov B.E., Taylor K.J. and Domracheva Y. (2008a) Stability
of cubic zirconia in a granitic system under high pressure and temperature,
Scientific Basis for Nuclear Waste Management, Mater. Res. Soc. Symp. Proc.
1107, 59–66.

Gibb F.G.F., Taylor K.J. and Burakov B.E. (2008b) The "granite encapsulation"
route to the safe disposal of Pu and other actinides, *J. Nucl. Mater.* **374**,
364–369.

Hambley M.J., Dumbill S., Maddrell E.R. and Scales C.R. (2008) Characterisation
of 20 year old [238]Pu-doped Synroc C, *Scientific Basis for Nuclear Waste
Management XXXI, Mater. Res. Soc. Symp. Proc.* **1107**, 373–380.

Hanchar J.M., Burakov B.E., Anderson E.B. and Zamoryanskaya M.V. (2003)
Investigation of single crystal zircon, (Zr,Pu)SiO$_4$, doped with [238]Pu, *Scien-
tific Basis for Nuclear Waste Management XXVI, Mater. Res. Soc. Symp.
Proc.* **757**, 215–225.

Hanchar J.M., Burakov B.E., Zamoryanskaya M.V., Garbuzov V.M., Kitsay A.A.
and Zirlin V.A. (2004) Investigation of Pu incorporated into zircon single
crystal, *Scientific Basis for Nuclear Waste Management XXVIII, Mater. Res.
Soc. Symp. Proc.* **824**, 225–236.

Hart K.P., Zhang Y., Loi E., Aly Z., Stewart M.W., Brownscombre A., Ebbinghaus B.B. and Bourcier W. (2000) *Scientific Basis for Nuclear Waste Management XXIII, Mater. Res. Soc. Symp. Proc.* **608**, 353–358.

Hatch L.P. (1953) Ultimate disposal of radioactive wastes, *Am. Scientist* **41**,
410–421.

Hyatt N.C. (2009) Private communication.

Heimann R.B. and Vandergraaf T.T. (1988) Cubic zirconia as a candidate waste
form for actinides: Dissolution studies, *J. Mat. Sci. Lett.* **7**, 583–586.

IAEA 257. (1985) *Chemical durability and related properties of solidified HLW
Forms*, Technical Reports Series No. 257, IAEA, Vienna.

Inoue M., Nishikawa T., Otsu H., Kominami H. and Inui T. (1998) Synthesis of
rare-earth gallium garnets by glycothermal method, *J. Am. Ceram. Soc.* **81**,
1173–1183.

Kanno Y. (1989) Thermodynamic and crystallographic discussion of the formation and dissociation of zircon, *J. Mater. Sci.* **24**, 2415–2420.

Kinoshita H., Kuramoto K-I., Uno M., Yamanaka S., Mitamura H. and Banba T.
(1998) Phase stability and mechanical property of yttria-stabilized zirconia

form for partitioned TRU wastes, *Proc. 2nd NUCEF Int. Symp. NUCEF'98*, Hitachinaka, Ibaraki, Japan, JAERI-Conf.99-004 (Part I), pp. 307–326.

Kinoshita H., Kuramoto K., Uno M., Yamanaka S., Mitamura H. and Banba T. (2000) Chemical durability of yttria-stabilized zirconia for highly concentrated TRU wastes, *Scientific Basis for Nuclear Waste Management XXIII, Mater. Res. Soc. Symp. Proc.* **608**, 393–398.

Kinoshita H., Kuramoto K., Uno M., Yamanaka S., Mitamura H. and Banba T. (2006) Mechanical integrity of yttria-stabilized zirconia doped with Np oxide, *Scientific Basis for Nuclear Waste Management XXIX, Mater. Res. Soc. Symp. Proc.* **932**, 647–654.

Kulyako Yu.M., Perevalov S.A., Vinokurov S.E., Myasoedov B.F., Petrov G.A., Ozhovan M.I., Dmitriev S.A. and Sobolev I.A. (2001) Properties of host matrices with incorporated U and Pu oxides, prepared by melting of a zircon-containing heterogeneous mixture (by virtue of exo effect of burning metallic fuel), *Radiochemistry* **43**, 626–631.

Kuramoto K-I., Makino Y., Yanagi T., Muraoka S. and Ito Y. (1995) Development of zirconia- and alumina-based ceramic waste forms for high concentrated TRU elements, *Proc. Int. Conf. GLOBAL'95*, Versailles, France **2**, 1838–1845.

Laverov N.P., Sobolev I.A., Stefanovsky S.V., Yudintsev S.V., Omelianenko B.I. and Nikonov B.S. (1998) Synthetic murataite — A new mineral host phase for immobilisation of actinides, *Reports of Russian Academy of Sciences* **362**(5), 670–672 (in Russian).

Laverov N.P., Velichkin V.I., Omelyanenko B.I., Yudintsev S.V., Petrov V.A. and Bichkov A.V. (2008) *Isolation of Spent Nuclear Materials: Geological and Geochemical Aspects* IGEM, RASci Moscow. (in Russian).

Lukinykh A.N., Tomilin S.V., Lizin A.A. and Bychkov A.V. (2002) Investigation of radiation and chemical stability of titanate ceramics intended for actinides disposal (B501111), in *Review of Excess Weapons Disposition: LLNL Contract Work in Russia*, eds. L.J. Jardine, G.B. Borisov, Proc. 3rd Annual Meet. for Coordination and Review of LLNL Work, St. Petersburg, Russia, (2002), UCRL-ID-149341, pp. 273–283.

Lukinykh A.N., Tomilin C.V., Lizin A.A. and Livshits T.C. (2008) Radiation and chemical durability of artificial ceramic based on ferrite garnet, *Radiokhimia* **50**(4), 375–379 (in Russian).

Lutze W. and Ewing R.C. (Eds.) (1988) *Radioactive Waste Forms for the Future*, (North-Holland Physics Publishing, the Netherlands).

Management and Disposition of Excess Weapons Plutonium. (1994) National Academy of Sciences (W.K.H. Panofsy, Study Chair), Committee of International Security and Arms Control, National Academy Press, Washington, D.C.

McCarthy G.J., White W.B. and Pfoertsch D.E. (1978) Synthesis of nuclear waste monazites, ideal actinide hosts for geological disposal, *Mater. Res. Bull.* **13**, 1239–1245.

McCarthy G.J. (1979) High level waste ceramics, materials considerations, process simulation and product characterisation, *Nuclear Technol.* **32**, 92.

106   *Crystalline Materials for Actinide Immobilisation*

McKown H.S., Smith D.H., Eby R.E. and Christie W.H. (1982) Differential lead retention in zircons: Implications for nuclear waste containment. *Science* **216**, 296–298.

Metcalfe B.L., Donald I.W., Scheele R.D. and Strachan D.M. (2003) Preparation and characterization of a phosphate ceramic for the immobilization of chloride-containing intermediate level waste, *Scientific Basis for Nuclear Waste Management XXVI, Mater. Res. Soc. Symp. Proc.* **757**, 265–271.

Metcalfe B.L., Donald I.W., Scheele R.D. and Strachan D.M. (2004) The immobilization of chloride-containing actinide waste in a calcium phosphate ceramic host: Ageing studies, *Scientific Basis for Nuclear Waste Management XXVIII, Mater. Res. Soc. Symp. Proc.* **824**, 255–260.

Morgan P.E.D. and Ryerson F.G. (1982) A new "cubic" crystal compound, *J. Mater. Sci. Lett.* **1**, 351–352.

O'Holleran T.P., Johnson S.G., Frank S.M., Meyer M.K., Noy M., Wood E.L., Knecht D.A., Vinjamuri K. and Staples B.A. (1997) Glass-ceramic waste forms for immobilizing plutonium, *Scientific Basis for Nuclear Waste Management XX, Mater. Res. Soc. Symp. Proc.* **465**, 1251–1258.

Ojovan M.I., Petrov G.A., Stefanovsky S.V. and Nikonov B.S. (1999) Processing of large scale radwaste-containing blocks using exothermic metallic mixtures, *Scientific Basis for Nuclear Waste Management XXII, Mater. Res. Soc. Symp. Proc.* **556**, 239–245.

Ojovan M.I., Karlina O.K., Klimov V.L., Trusov B.G., Pavlova G.Y. and Sobolev I.A. (2001) Self-sustaining reactions for the processing technologies of chemically stable matrices incorporating carbon and zirconium wastes, *Proc. 8th Int. Conf. Rad. Waste Manag. Envir. Remed*, Bruges, Belgium. 30 September to 4 October, 2001; CD-ROM. Paper 83, ASME.

Ojovan M.I. and Lee W.E. (2007) *New Developments in Glassy Nuclear Wasteforms* (Nova Science Publishers, New York).

Pena P. and De Aza S. (1984) The zircon thermal behaviour: Effect of impurities. Part 1. *J. Mater. Sci.* **19**, 135–142.

Poirot I., Kot W., Shalimoff G., Edelstein N., Abraham M.M., Finch C.B. and Boatner L.A. (1988) Optical and EPR investigations of Np^{4+} in single crystals of $ZrSiO_4$, *Phys. Rev.* B **37**, 3255–3264.

Poirot I., Kot W.K., Edelstein N.M., Abraham M.M., Finch C.B. and Boatner L.A. (1989) Optical study and analysis of Pu^{4+} in single crystals of $ZrSiO_4$, *Phys. Rev.* B **39**(10), 6388–6394.

Raison P.E., Haire R.G., Sato T. and Ogawa T. (1999) Fundamental and technological aspects of actinide oxide pyrochlores: Relevance for immobilization matrices, *Scientific Basis for Nuclear Waste Management XXII, Mater. Res. Soc. Symp. Proc.* **556**, 3–10.

Raison P.E., Haire R.G. and Assefa Z. (2002) Fundamental aspects of Am and Cm in zirconia-based materials: Investigations using X-ray diffraction and Raman spectroscopy, *J. Nucl. Sci. Tech.*, Suppl. **3**, 725–728.

Richman I., Kisliuk P. and Wong E.Y. (1967) Absorption spectrum of U^{4+} in zircon ($ZrSiO_4$), *Phys. Rev.* **155**(2), 262–267.

Ringwood A.E. (1978) *Safe Disposal of High-Level Nuclear Reactor Wastes: A New Strategy*, (Australian National University Press, Canberra, Australia).

Ringwood A.E., Kesson S.E., Reeve K.D., Levins D.M. and Ramm E.J. (1988) in *Radioactive Waste Forms for the Future*, eds. W. Lutze and R.C. Ewing, (North-Holland Physics Publishing, the Netherlands), pp. 233–334.

Roy R. (1975) Ceramic science of nuclear waste fixation, *Am. Ceram. Soc. Bull.* **54**, 459.

Seida Y., Yuki M., Suzuki K. and Sawa T. (2003) Sodium zirconium phosphate [NZP] as a host matrix for high level radioactive waste, *Scientific Basis for Nuclear Waste Management XXVI, Mater. Res. Soc. Symp. Proc.* **757**, 329–334.

Sikora R.E., Raison P.E. and Haire R.G. (2005) Self-irradiation induced structural changes in the transplutonium pyrochlores $An_2Zr_2O_7$ (An = Am, Cf), *J. Solid State Chemistry* **178**, 578–583.

Sobolev I.A., Stifanovsky S.V., Youdintsev S.V., Nikonov B.S., Omelianenko B.I. and Mokhov A.V. (1997) Study of melted Synroc doped with simulated high-level waste, *Scientific Basis for Nuclear Waste Management XX, Mater. Res. Soc. Symp. Proc.* **465**, 363–370.

Sobolev I.A., Ojovan M.I., Petrov G.A., Klimov V.L. and Tarasov V.L. (1998) Self-sustaining synthesis of Synroc: Thermodynamic analysis, *Proc. Conf. Incineration and Thermal Treatment Technologies*, Salt Lake City, Utah, USA, 11–15 May 1998, pp. 311–313.

Stefanovsky S.V., Yudintsev S.V., Nikonov B.S., Omelianenko B.I. and Ptashkin A.G. (1999) Murataite-based ceramics for actinide waste immobilization, *Scientific Basis for Nuclear Waste Management XXII, Mater. Res. Soc. Symp. Proc.* **556**, 121–128.

Stefanovsky S.V., Kiryanova O.I., Yudintsev S.V., Nikonov B.S. and Omelianenko B.I. (2001) Phase composition and chemical elements distribution in murataite-based ceramics containing rare-earths and actinides, *Physics and Chemistry of Material Treatment* **3**, 72–80 (in Russian).

Stefanovsky S.V., Yudintsev S.V., Gieré R., Lumpkin G.R. (2004) Nuclear waste forms, in *Energy, Waste, and the Environment: A Geochemical Perspective*, Eds. R. Gieré and P. Stille, Geol. Soc., London, Special Publ., **236**, 37–63.

Strachan D.M., Scheele R.D., Buchmiller W.C., Vienna J.D., Sell R.L. and Elovich R.J. (2000) Preparation of ^{238}Pu-ceramics for radiation damage experiments, Report PNNL-13251 under contract DE-A-C06-76RLO 1830.

Strachan D.M. (2001) Glass dissolution: Testing and modeling for long-term behavior, *J. Nucl. Mater.* **298**, 69–77.

Strachan D.M., Scheele R.D., Kozelisky A.E., Sell R.L., Todd Schael H., O'Hara M.J., Brown C.F. and Buchmiller W.C. (2002) Radiation damage in titanate ceramics for plutonium immobilization, *Scientific Basis for Nuclear Waste Management XXV, Mater. Res. Soc. Symp. Proc.* **713**, 461–468.

108   *Crystalline Materials for Actinide Immobilisation*

Strachan D.M., Scheele R.D., Icenhower J.P., Buck E.C., Kozelisky A.E., Sell
    R.L., Elovich R.J. and Buchmiller W.C. (2004) Radiation damage effects in
    candidate ceramics for plutonium immobilization: Final Report, PNNL-14588,
    Richland, Washington 99352.
Strachan D.M., Scheele R.D., Buck E.C., Icenhower J.P., Kozelisky A.E., Sell
    R.L., Elovich R.J. and Buchmiller W.C. (2005) Radiation damage effects in
    candidate titanates for Pu disposition: Pyrochlore, *J. Nucl. Mater.* **345**(2–3),
    109–135.
Strachan D.M., Scheele R.D., Buck E.C., Kozelisky A.E., Sell R.L., Elovich R.J.
    and Buchmiller W.C. (2008) Radiation damage effects in candidate titanates
    for Pu disposition: Zirconolite, *J. Nucl. Mater.* **372**, 16–31.
Timofeeva L.F., Nadykto B.A., Orlov V.K., Malyukov E.E., Molomin V.I., Zhmak
    V.A., Semova E.A., Shishkov N.V. (2002) Preparation and study of the
    critical-mass-free plutonium ceramics with neutron poisons Hf, Gd and Li,
    J. Nucl. Sci. and Technology Suppl, **3**, 729–732.
Volkov Yu.F., Lukinykh A.N., Tomilin S.V. and Bychkov A.V. (2001) Investiga-
    tion of U.S. titanate ceramics radiation damage due to ^{238}Pu alpha-decay
    [B501111], in *Immobilization of Excess Weapons Plutonium in Russia: A
    Review of LLNL Contract Work*, eds. L.J. Jardine, G.B. Borisov, Proc. Meet.
    for Coordination and Review of Work, St. Petersburg, Russia, (2000), UCRL-
    ID-143846, pp. 196–207.
Volkov Yu.F., Tomilin S.V., Orlova A.I., Lizin A.A., Spiryakov V.I. and Lukinykh
    A.N. (2003) Phosphates of actinides $AM_2(PO_4)_3$ (M^{IV} = U, Np, Pu; A^I = Na,
    K, Rb) with rhombohedral structure, *Radiokhimia* **45**, 289–297 (in Russian).
Wang S.X., Begg B.D., Wang L.M., Ewing R.C., Weber W.J. and Govidan Kutty
    K.V. (1999) Radiation stability of gadolinium zirconate: A waste form for
    plutonium disposition, *J. Nucl. Mater.* **14**, 4470–4473.
Weber W.J., Turcotte R.P., Bunnell L.R., Roberts F.P. and Westsik J.H. (1979)
    Radiation effects in vitreous and devitrified simulated waste glass, in *Ceram-
    ics in Nuclear Waste Management*, eds. T.D. Chikalla and J.E. Mendel,
    CONF-790420, Nationals Technical Information Service, Springfield, Virginia,
    pp. 294–299.
Weber W.J. (1982) Radiation damage in rare-earth silicate with the apatite struc-
    ture, *J. Am. Ceram. Soc.* **65**, 544–548.
Weber W.J. (1983) Radiation-induced swelling and amorphization Ca_2Nd_8
    $(SiO_4)_6O_2$, *Radiation Effects* **77**, 295–308.
Weber W.J., Wald J.W. and Matzke Hj. (1986) Effect of self-radiation damage
    in Cm-doped $Gd_2Ti_2O_7$ and $CaZrTi_2O_7$, *J. Nucl. Mater.* **138**, 196–209.
Yudintsev S.V., Stefanovsky S.V. and Ewing R.C. (2007) Actinide host phases
    as radioactive waste forms, in *Structural Chemistry of Inorganic Actinide
    Compounds.*, eds. S. Krivovichev, P. Burns, and I. Tananaev (Elsevier BVP),
    pp. 453–490.
Yudintsev S.V., Lukinykh A.N., Tomilin S.V., Lizin A.A. and Stefanovsky S.V.
    (2009) Alpha-decay induced amorphization in Cm-doped Gd_2TiZrO_7, *J. Nucl.
    Mater.* **385**, 200–203.

Zamoryanskaya M.V., Burakov B.E., Bogdanov R.R. and Sergeev A.S. (2002)
A cathodoluminescence investigation of pyrochlore, $(Ca,Gd,Hf,U,Pu)_2Ti_2O_7$,
doped with ^{238}Pu and ^{239}Pu, *Scientific Basis for Nuclear Waste Management
XXV, Mater. Res. Soc. Symp. Proc.* **713**, 481–485.

Zamoryanskaya M.V. and Burakov B.E. (2004) Electron microprobe investigation of Ti-pyrochlore doped with Pu-238, *Scientific Basis for Nuclear Waste
Management XXVIII, Mater. Res. Soc. Symp. Proc.* **824**, 231–236.

Chapter 4

Synthesis Methods

4.1. Precursor Fabrication

The starting precursors play an important role in further synthesis of actinide-doped ceramics. Such ceramics should not contain separate actinide phases so the main aim of precursor fabrication is to give homogeneous and chemically-active starting materials, which enable complete actinide incorporation into the crystal structures of host phases, often in the form of solid solutions during ceramic fabrication.

4.1.1. *Sol-gel*

A sol is a highly dispersed arrangement of 1–1,000 nm colloid particles in liquid (Lee and Rainforth, 1994). Increasing the volume of particles, removal of some liquid or use of polymerisation reactions causes formation of linking bonds between them and gelation. Gelation may take place at dispersed phase (particles) content from 0.1 to several wt.%. The term sol-gel includes manufacture of products made from both inorganic colloidal particles suspended in aqueous solutions (particulate systems) and via alkoxides which can be partially hydrolysed and then polymerised into a gel (polymeric systems) where a classical sol never existed (Lee and Rainforth, 1994). Sol-gel processing was developed in the 1950s for fabrication of powders of UO_2

and ThO_2 for nuclear fuels without production of large quantities of hazardous dust. The gelation of liquid HLW or non-radioactive solutions is usually initiated by so called gel-forming agents such as tetraethoxysilane (TEOS), $Si(OC_2H_5)_4$.

Actinide-doped zircon/zirconia ceramics have been synthesised using TEOS alkoxide as the gel-forming agent as well as the source of Si. The procedure of gelation appears very simple, however there are important experimental details that need to be taken into account:

(a) At the beginning two solutions are prepared separately: nitrates or chlorides of actinides and Zr are dissolved in water, but TEOS should first be dissolved in acetone or methanol and then water can be added. Direct mixing of TEOS and water causes rapid hydrolysis of $Si(OC_2H_5)_4$ and precipitation of hydrated silica;

(b) Nitrate solution and TEOS (water-acetone or water-methanol solution) are mixed together while stirring to give a completely homogeneous single-phase transparent solution;

(c) pH of the common solution should be controlled. Decreasing pH causes gelation. However, adding excessive acid may cause overly rapid gelation and lead to inhomogeneity in the solidified gel.

Newly solidified gel is transparent and rubber-like (Fig. 4.1.1). Slow drying at room temperature causes complete gel solidification to the brittle glass-like state. Usually, this process is accompanied by crack formation (Fig. 4.1.2) although it is possible to preserve solid crack-free gel (Brinker and Scherer, 1990).

Further calcination from 300 to $1,000°C$ (Fig. 4.1.3) causes complete release of water, organics and nitrogen oxide from the gel matrix. However, the matrix of calcined material remains generally amorphous with some unidentified crystallites (Fig. 4.1.4). Sintering of the calcined gel at $1,000–1,100°C$ causes growth of crystalline zirconia spheroids in an amorphous silica matrix (Fig. 4.1.5).

In experiments with Ce and Gd admixtures as actinide surrogates zirconia spheroids accommodate all the dopants keeping the amorphous silica matrix free of Ce and so presumably U. This observation

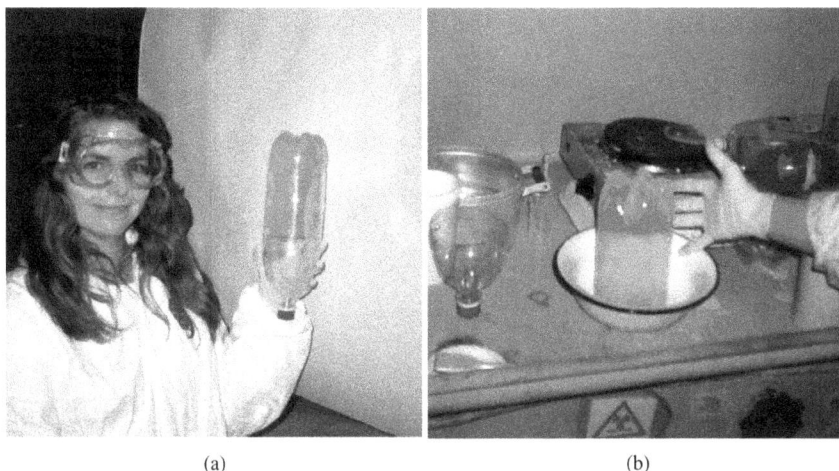

Fig. 4.1.1. (a) U-doped Zr-silicate gel immediately after solidification from aqueous solution; (b) removed from the plastic bottle the gel retains its initial shape.

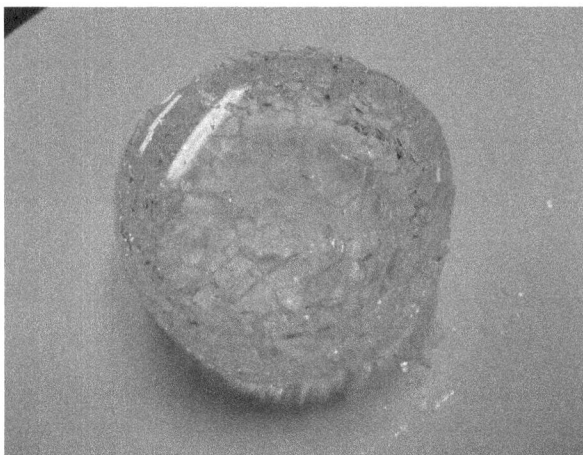

Fig. 4.1.2. Crack formation in Zr-silicate gel as a result of drying for two days at 25°C in air.

explains the unusually high chemical durability of natural uranium-bearing "gel-zircon" (see Sec. 1.4.14) and artificial Pu-Am-doped Zr-silicate gel (Burakov *et al.*, 2006). Incorporation of actinides into the structure of zirconia crystallites starts in the gel matrix.

Fig. 4.1.3.   U-doped Zr-silicate gel after calcination for 1 hour in air at 350°C.

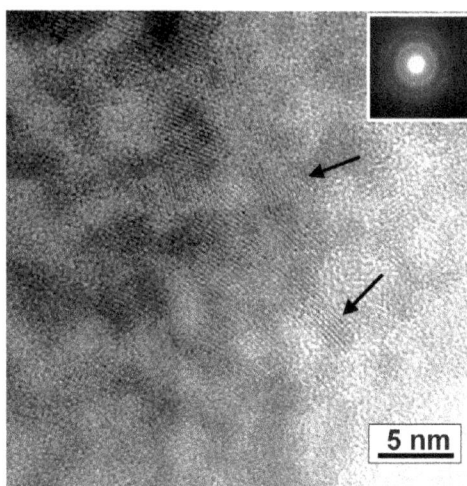

Fig. 4.1.4.   High resolution TEM image of Ce-doped Zr-silicate gel calcined for 1 hour in air at 600°C. Unidentified crystallites (probably zirconia) are marked by arrows. Electron diffraction (insert) indicates amorphous matrix.

Separate actinide phases do not form during gel formation, transformation, solidification and crystallisation to actinide-doped zircon.

Formation of zircon from sol-gel precursors (Ushakov *et al.*, 1998) begins after 1 hour at 1,400°C. However, using other chemicals for sol-gel preparation and/or longer calcination and sintering times may allow zircon formation at lower temperatures.

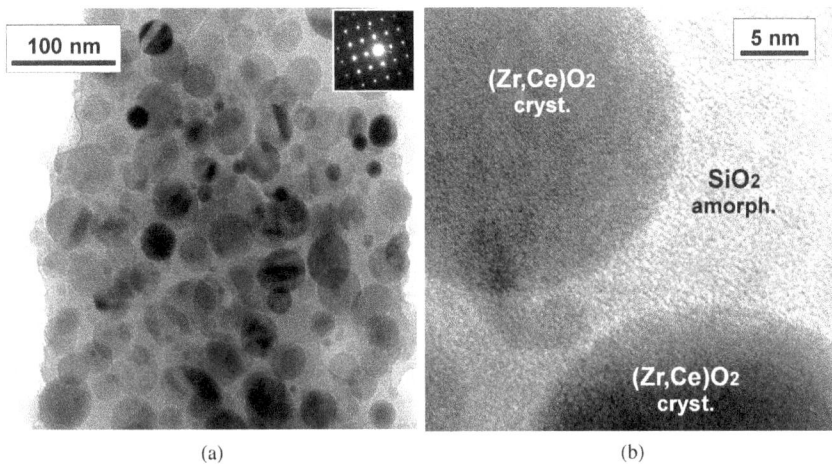

Fig. 4.1.5.   TEM images at low (a) and high (b) resolution of Ce-doped Zr-silicate gel sintered 1 hour at 1,300°C. Electron diffraction (insert in (a)) indicates tetragonal zirconia spheroids $(Zr,Ce)O_2$. The Ce is incorporated into the crystalline structure of tetragonal zirconia. Amorphous silica matrix is free of Ce.

4.1.2. *Co-precipitation*

The co-precipitation method (Lee and Rainforth, 1994) is based on simple conversion of actinides and non-radioactive elements from common aqueous solutions into insoluble hydroxide, oxalate or phosphate powders. The main advantage of this method is that it avoids formation of separate actinide phases during co-precipitation. All actinides should form solid solutions with the precipitating solid phase. In some cases co-precipitation may already cause formation of partly crystalline host phases such as fluorapatite, monazite, thorium phosphate-diphosphate (TPD). This process might be accompanied by stabilisation of unusual actinide valence states. For example, Pu^{3+} has been observed in monazite powders, $(La,Pu)PO_4$ and $PuPO_4$, precipitated from nitrate solutions in air (Burakov *et al.*, 2004).

The use of high volumes of aqueous solutions and the necessity to filter and dry large volumes of precipitates is a disadvantage of the co-precipitation method.

116   *Crystalline Materials for Actinide Immobilisation*

Some important details of co-precipitation include:

(a) Precipitation of actinide-doped monazite, (Ce,Gd,Eu,La,An) PO_4, from nitrate solution on adding H_3PO_4 may be very slow. In such cases the use of excess ammonium hydroxide is needed to provide full precipitation;
(b) To support homogeneous co-precipitation the actinide-bearing solutions are added to excess oxalate solution, ammonium hydroxides, etc., while stirring. Adding precipitating chemicals (oxalate and hydroxide solutions) to actinide-bearing solutions may cause a multi-phase precipitate to form.

Precursors obtained by co-precipitation have been used for synthesis of actinide-doped ceramics based on cubic zirconia, monazite, garnet and chlorapatite (see Chap. 3).

4.1.3. *Oxide powder mix*

Mixing of oxide powders is a typical inexpensive commercial method for ceramic production (Lee and Rainforth, 1994). Commercially-available oxides of Al, Zr, Si, Mg, etc., are mixed and ground together in different proportions depending on the desired ceramic composition. Ball-mills (Fig. 4.1.6) are commonly used for grinding oxides at laboratory and industrial scales to particle sizes down to a few

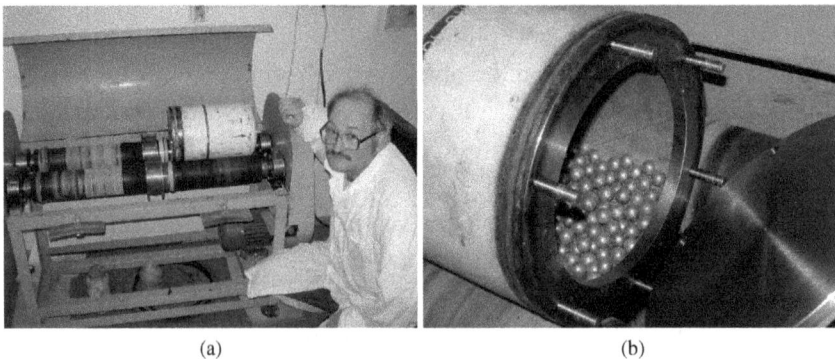

(a)   (b)

Fig. 4.1.6.   (a) Steel ball-mill for grinding non-radioactive oxide powders: general view and (b) opened milling canister.

microns. They are made of steel, ceramic or plastic depending on the necessity to keep the powders free of wear contamination.

To make sub-micron powders, special energy intensive mills are needed such as vibratory mills, which fracture particles by impact or attrition wear mechanisms (Lee and Rainforth, 1994).

A disadvantage of the mixed oxide route for synthesis of actinide-doped ceramics is incomplete reaction between actinide oxides and non-radioactive precursor components leading to actinide oxide inclusions in the final ceramic. Also, some non-radioactive oxides such as ZrO_2 and Al_2O_3, as well as calcined PuO_2 and NpO_2, are chemically inert. Chemical "reactivity" of oxide precursors can be increased using:

(a) Specially prepared amorphous and/or highly dispersed Al_2O_3; SiO_2; PuO_2, etc.;
(b) Admixtures of metallic powders (Zr, Ti, Al) or even substituting completely some inert oxides, for example, ZrO_2 for metallic Zr;
(c) Mechanical activation of powders through special high-energy milling (Sepelak, 2002; Batyukhnova and Ojovan, 2009);
(d) Impregnation of actinide solutions into non-radioactive oxide powders followed by further calcining and milling;
(e) Higher temperature, pressure and longer duration of ceramic synthesis.

The mixed oxide method has been used to synthesise, e.g., actinide-doped Synroc, Ti-pyrochlore, zirconolite and cubic zirconia (see Chap. 3).

4.2. Hot Uniaxial Pressing (HUP)

Hot uniaxial pressing (HUP) is a method of ceramic synthesis based on simultaneous heating and die pressing a green ceramic body. HUP in a graphite or metallic die is usually used for laboratory-scale experiments. Large samples of Synroc (30–60 kg and 300–400 mm in diameter) were successfully obtained in metallic bellows at $1,100$–$1,150°C$ and 14–21 MPa (Ringwood *et al.*, 1988).

HUP gives dense ceramics but needs development of special complex equipment and there are strict radiation safety requirements for handling actinides at high pressure. Current progress on successful synthesis of dense actinide-doped ceramics at atmospheric pressure (see Sec. 4.4) has decreased the attractiveness of HUP for industrial application in this field.

4.3. Hot Isostatic Pressing (HIP)

Hot isostatic pressing (HIP) simultaneously applies heat and isostatic pressure to a green compact in a sealed chamber (Lee and Rainforth, 1994). Isostatic pressing applies even pressure via a flexible mould, which is inserted in a vessel of fluid to which pressure is applied. HIP may provide much higher and more even pressures compared to HUP and allows production of dense polycrystalline materials, e.g., PuO_2 with greater than 99% of theoretical density. It is a potential option to synthesise ceramic wasteforms for immobilisation of waste derived from spent MOX fuel (Maddrell and Abraitis, 2004). However, HIP is unlikely to be used commercially due to its high cost and the problem of licensing for treatment of actinides under high pressure in a hot gas medium.

4.4. Pressing-sintering

Cold pressing followed by sintering is a popular method of laboratory-scale ceramic synthesis. Lawrence Livermore National Laboratory (LLNL) has successfully used it for the fabrication of full-size samples (approximately 6 cm in diameter) of Pu-doped pyrochlore ceramic (see Chap. 3). Its main advantage is the use of relatively inexpensive and compact equipment, which is easily installed into a glove box. Ceramic powder is loaded into a steel die (Fig. 4.4.1), cold uniaxially pressed into pellets (Fig. 4.4.2) and then these pellets are sintered in a furnace (Fig. 4.4.3).

There are two problems with this method related to extraction of pressed pellets from a die and control of heating rate on sintering.

Fig. 4.4.1.   Manual loading of steel die of non-radioactive precursor at the Immobilisation Science Laboratory, Sheffield, UK.

(a)   (b)

Fig. 4.4.2.   Manual cold pressing of non-radioactive precursor at the Immobilisation Science Laboratory, Sheffield, UK.

Precursor powders may be pressed without any binder although this can impact on the final sintered product density. Optimal pressure should be verified experimentally and separately for each type of powder. Pressures of about 2–5 MPa are used and never above 20 MPa. Excessive pressure may cause pellets to crack on extraction from the die, but low pressure might cause high ceramic porosity. To

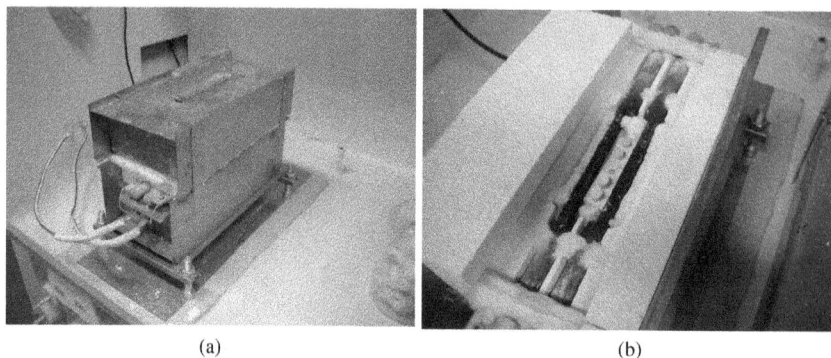

(a)   (b)

Fig. 4.4.3.   Special compact furnace developed by KRI for synthesis of actinide-doped ceramics in a glove box at temperature up to 1,600°C in air: (a) general view; (b) with open lid. Furnace body is made up of porous alumina bricks which prevent release of radioactive aerosols from the high-temperature core. Four ceramic pellets between two exchangeable SiC heating elements can be seen in (b).

optimise the cold pressing process, organic binders are often used — for example, an aqueous solution of polyvynyl-alcohol (PVA). Small amounts of liquid binder are added to the precursor and this mixture is then ground in a ball-mill or mortar until it becomes homogeneous and dry. The binder acts as a glue for precursor particles and at the same time as a lubricant, supporting easy extraction of crack-free pellets from the die. For each type of precursor, the optimal amount of binder is determined experimentally. An excessive amount of binder may cause high porosity in the fired ceramic as it burns out. For the same precursor the optimal amount of binder for small and large pellets might be different. Binder evaporation takes place on heating at 200–300°C and it is important to provide a slow enough heating rate during this range for complete binder evaporation (Richerson, 1992).

4.5.   Melting-crystallisation

Synthesis of ceramics by melting and crystallisation processes is used extensively, e.g., for production of glass ceramics and fused $(Al_2O_3\text{-}ZrO_2\text{-}SiO_2)$ refractory bricks. Some actinide host phases such

as zirconolite, Ti-pyrochlore, garnet, perovskite, murataite, fluorapatite, silicate apatite etc., could be synthesised from melted precursors at 1,400–2,000°C. Reasons for considering the use of melting
technology for synthesis of actinide-doped ceramics include:

(a) Recent development of CCM technology based on induction
    heating to 3,000–3,500°C (Fig. 4.5.1), which allows production of
    large (10-cm size) polycrystalline samples (Fig. 3.5.4) and single
    crystals (Fig. 4.7.1);
(b) The complex chemical and phase composition of some actinide
    wastes. Preparation of homogeneous powdered precursors from
    actinide-bearing residues for further synthesis of ceramic waste
    forms by sintering method is difficult and often not economi-
    cally viable. Melting avoids the need to mill and cold press. The
    use of a flexible multi-phase ceramic composition may overcome

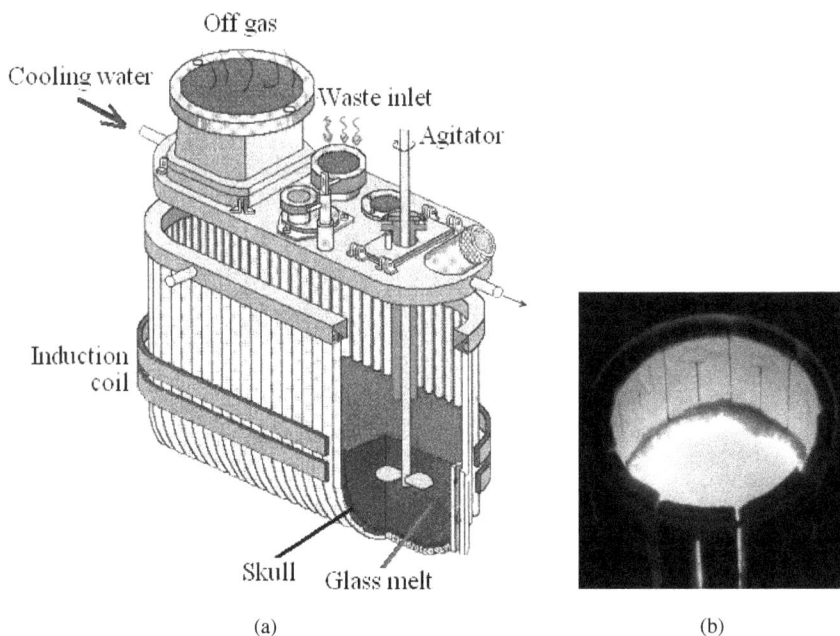

(a)   (b)

Fig. 4.5.1.   (a) Schematic of a cold crucible melter (Ojovan and Lee, 2007), and
(b) melt surface in CCM above 2,000°C, courtesy of Dr. B.T. Melekh of the Ioffe
Physical-Technical Institute, St. Petersburg, Russia.

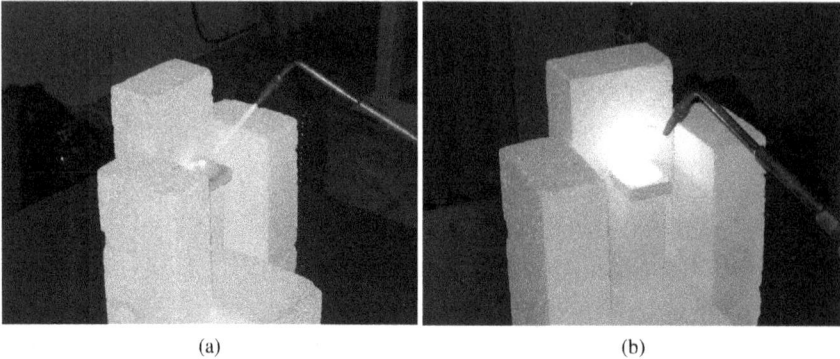

(a)   (b)

Fig. 4.5.2.   Synthesis of small samples of aluminate garnet by melting in a hydrogen flame above 2,000°C. High-temperature area is shielded by porous alumina
bricks.

the problem of precursor inhomogeneity. Synroc samples have
been successfully synthesised by melting-crystallisation (Sobolev
et al., 1997, Xu and Wang, 2000).

Basic synthesis of actinide-doped samples by melting could be
carried out using a simple hydrogen torch reaching 2,500–2,800°C
(Fig. 4.5.2). This equipment is straightforward to install into a glove
box and Pu-doped garnet/perovskite ceramics have been made via
this method (Figs. 3.5.1 and 3.5.3).

4.6. Self-sustaining (Self-propagating) High Temperature Reactions

Ceramic processing using self-propagating high-temperature synthesis is based on exothermic reactions between reactive metallic
powders such as Ti, Zr and Al, and oxides such as MoO_3 and
Fe_2O_3. The reactions are accompanied by release of heat and green
compacts achieve temperatures of 1,700–2,500°C for several sec-
onds in actinide-containing systems (Glagovskiy *et al.*, 2001). Control of the oxide/metal ratio allows the reactions to be controlled,
resulting in formation of large (10-cm size) monolithic ceramic
samples (Fig. 4.6.1) (Ojovan *et al.*, 1999; Ojovan and Lee, 2007).

Fig. 4.6.1. Self-sustaining production of large monolithic samples (25 kg) in a double-wall container crucible.

Yields of ~15–20 wt.% have been achieved from self-sustaining reaction of precursor mixtures of SiO_2 and Zr powder. Adding U_3O_8 to the precursor allows production in air of U-doped zircon, $(Zr,U)SiO_4$, and zirconia, $(Zr,U)O_2$ (Fig. 4.6.2). Note that doping of any host phase with U^{4+} using other synthesis methods in air is very difficult due to uranium oxidation.

Even though self-sustaining synthesis is not well developed in actinide materials some samples of Ti-pyrochlore ceramic doped with Pu (10–30 wt.% PuO_2), Np (10–30 wt.% NpO_2) and Am (0.5 wt.% Am_2O_3) have been obtained (Yudintsev *et al.*, 2004). Some polyphase zircon-based glass composite materials have also been obtained to immobilise U (1–12 wt.% U_3O_8) and Pu (1 wt.% PuO_2) (Kulyako *et al.*, 2001).

4.7. Single Crystal Growth

Synthesis of actinide-doped crystals is important in a number of areas, including actinide spectroscopy, standards for actinide analysis

124   *Crystalline Materials for Actinide Immobilisation*

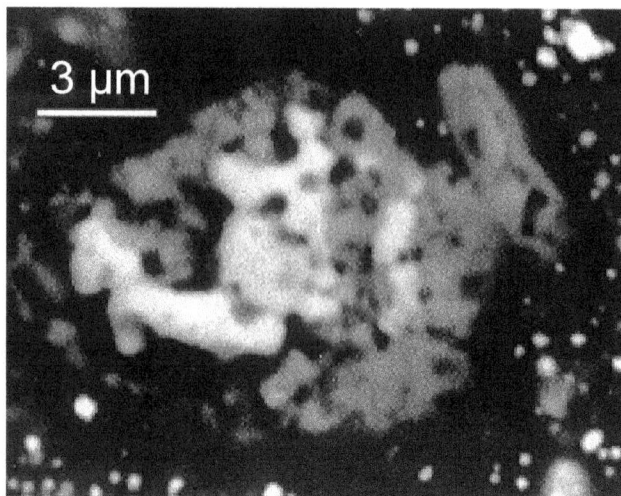

Fig. 4.6.2.   Backscattered electron SEM image of U-doped (up to $10\,wt.\%$ U) zircon and zirconia phases synthesised in air by self-propagating reaction in a precursor mixture of $U_3O_8 + SiO_2 +$ metal Zr. Light phase is zirconia, $(Zr,U)O_2$, and grey phase is zircon, $(Zr,U)SiO_4$. Black area is acrylic resin.

(e.g., EPMA), development of advanced materials (see Chap. 2) and detailed study of the main features of actinide hosts. Methods of crystal growth have been reported, e.g., Elwell (1979); Balitskiy and Lisitsina (1981). However, most of these methods have not been used to obtain highly radioactive crystals although there is industrial production of large, non-radioactive monocrystals of some prospective actinide hosts such as garnet ($Y_3Al_5O_{12}$, $Gd_3Ga_5O_{12}$, etc.) and cubic zirconia (Fig. 4.7.1) for use as laser and luminescence materials as well as artificial gemstones.

When synthesising crystals doped with, for example, Pu, Np and Am in a glove box it is necessary to avoid radioactive aerosol release. Also, it is sensible to minimise the amount of actinides produced, which might otherwise lead to secondary radioactive wastes.

Crystals of monazite doped with Pu, Np, Am, Cm have been obtained at ORNL (Boatner *et al.*, 1980) and PuO_2 (Fig. 1.3.4); NpO_2 (Fig. 1.3.5); Pu-doped zircon (Figs. 2.3.4. and 4.7.5); and Am-doped monazite (Fig. 4.7.5) obtained at KRI by the flux method.

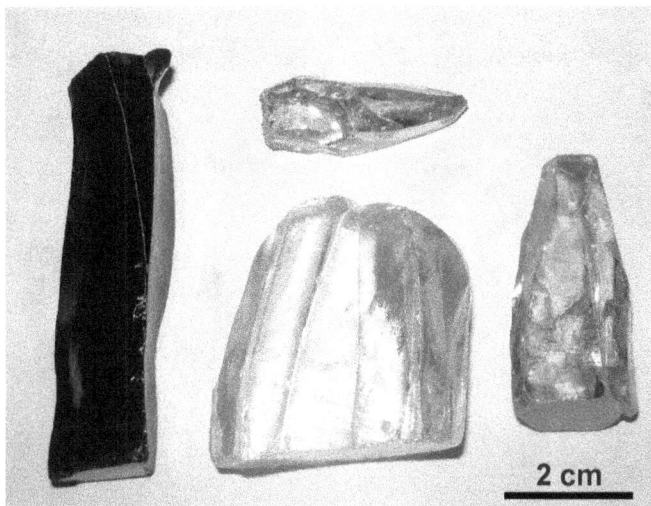

Fig. 4.7.1.   Single crystals of yttria-stabilised cubic zirconia, $(Zr,Y,\ldots)O_2$ grown in a CCM.

This method is based on reaction between actinide oxides and non-radioactive chemicals such as SiO_2, ZrO_2, phosphates of rare-earths dissolved in a melted flux salt such as MoO_3, Li_2MoO_4, $CaWO_4$ and LiF. Partial flux evaporation and/or slow cooling causes oversaturation of dissolved species in the melt and formation and crystallisation of the desired phases. Zircon crystal growth by the flux method has been described (Hanchar *et al.*, 2001; Hanchar *et al.*, 2004). A similar procedure has been used to make crystals of actinide-doped monazite, xenotime, zirconia and garnet (Anderson and Burakov, 2004; Hanchar *et al.*, 2003; Hanchar *et al.*, 2004 and Burakov *et al.*, 2009).

Important details of the flux method include:

(a) The furnace used in the glove box (Fig. 4.7.2) should be equipped with a precise temperature controller and be designed to work at up to $1,200°C$ for several days. A special system that prevents release of radioactive aerosols during flux evaporation is required (Kitsay *et al.*, 2004);

126   *Crystalline Materials for Actinide Immobilisation*

Fig. 4.7.2.   Tube furnace developed by KRI (Kitsay *et al.*, 2004) for synthesis of actinide-doped crystals by the flux method in air at up to 1,200°C. The furnace design allows accumulation of evaporating flux in a special internal cold zone that prevents radioactive aerosol release into the glove-box ventilation system.

(b) For high yield of large (mm-size), well-shaped crystals all crystal-forming components of the precursor should be calcined at 500–900°C and ground before being loaded into the crucible;

(c) In some cases it is difficult to provide careful mixing and grinding of the flux salt(s) and highly radioactive precursors before loading into the crucible. To overcome this problem it is possible to initially grind non-radioactive precursor components and flux salt(s) and then to add the actinide oxide (without complete mixing);

(d) The final blend of flux and precursor is placed in a lidded Pt-crucible, which limits flux evaporation;

(e) Under flux evaporation crystal formation takes place in different zones of the crucible (Fig. 4.7.3), but the best shaped and largest crystals are usually precipitated at the crucible bottom or on the walls near the bottom (Figs. 4.7.4 and 4.7.5);

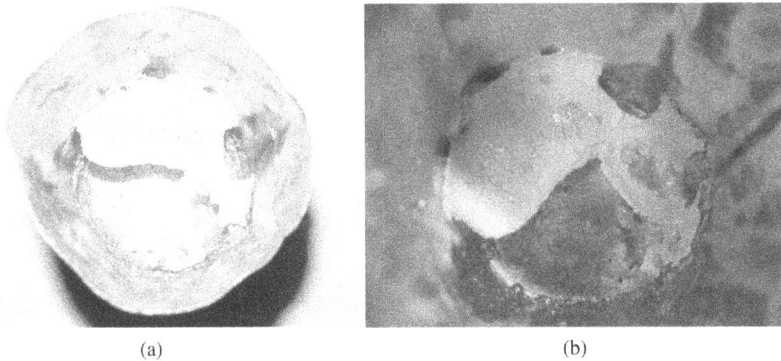

(a)   (b)

Fig. 4.7.3.   Platinum crucible immediately after growth of zircon single crystals doped with Pu, Eu and Tb: (a) general view and (b) close image of crucible bottom. There are three areas of crystal location: crucible bottom filled with the rest of flux (1); walls which are almost free of flux (2); on the surface of flux film (white) above crucible bottom (3).

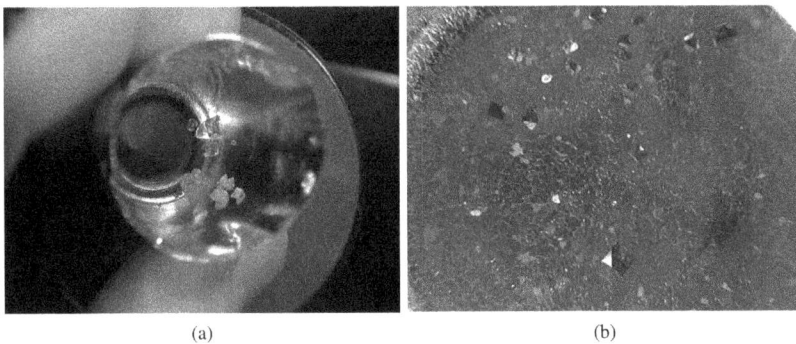

(a)   (b)

Fig. 4.7.4.   (a) Single crystals of V-doped zircon and (b) Pu-doped (8–14 wt.% Pu) zircon from bottom of Pt-crucible after dissolution of the rest of the flux.

(f) The highest flux evaporation temperature does not usually exceed 1,200°C. Crystal growth via flux evaporation occurs only at a constant temperature or on slow cooling;

(g) A combination of flux evaporation and slow cooling produces larger crystals. However, to homogeneously distribute the actinide admixture in some crystals (for example, zircon), it is necessary to evaporate the flux at fixed temperature for 2–5 days followed by rapid cooling.

Fig. 4.7.5.   (a) Single crystals of Pu-doped (8–14 wt.% [239]Pu) zircon, (Zr,Pu)SiO$_4$ (a) and (b) Am-doped (0.3 wt.% [243]Am) monazite, (Dy,La,Am)PO$_4$, obtained by flux method and crystallised in lower part of Pt-crucible.

4.8. Summary

Bulk chemical composition of actinide wastes strongly affects the choice of precursor and method of ceramic synthesis. The most homogeneous and chemically "reactive" precursors are prepared by sol-gel and co-precipitation methods although both methods require large volumes of radioactive solutions. The use of "dry" mixed oxide precursors is common in the ceramic industry but has some disadvantages for actinide ceramic wasteform production, including low precursor "reactivity" and formation of highly alpha-radioactive dust and aerosols.

Currently, synthesis of actinide-doped ceramics tends to be done at atmospheric pressure, avoiding application of HUP and HIP methods. Licensing of actinide treatment at high pressure is becoming more and more difficult. Use of pressing-sintering methods is common for laboratory-scale synthesis of actinide-doped ceramics and is likely to find industrial adoption in the near future. The melting-crystallisation method, in particular, using cold crucible melters for ceramic wasteform production has potential, but requires further development and testing. Self-sustaining (self-propagating) high-temperature synthesis might be applicable for immobilisation of actinide-bearing wastes of complex chemical composition. However, detailed research is needed to justify use of this method. Synthesis, study and application of actinide-doped single crystals are at an early stage of development despite the fact that industrial technologies for obtaining large non-radioactive monocrystals are well established.

References

Anderson E.B. and Burakov B.E. (2004) Ceramics for the immobilization of plutonium and americium: Current progress of R&D of the V.G. Khlopin Radium Institute, *Scientific Basis for Nuclear Waste Management XXVII, Mater. Res. Soc. Symp. Proc.* **807**, 207–212.

Balitskiy V.S. and Lisitsina E.E. (1981) *Synthetic Analogues and Imitations of Natural Gem Stones* (Moscow, "Nedra") (in Russian).

Batykhnova O.G. and Ojovan M.I. (2009) Tribiochemical treatment for immobilisation of radioactive wastes, *Scientific Basis for Nuclear Waste Management XXXII, Mater. Res. Soc. Symp. Proc.*, Vol. 1124, Warrendale, PA, 6 pp.

Boatner L.A., Beall G.W., Abraham M.M., Finch C.B., Huray P.G. and Rappaz M. (1980) Monazite and other lanthanide orthophosphates as alternative actinide waste forms, *Scientific Basis for Nuclear Waste Management*, **2**, 2899–296.

Brinker C.J. and Scherer G.W. (1990) *Sol-gel Science* (Academic Press, London).

Burakov B.E., Smetannikov A.Ph. and Anderson E.B. (2006) Investigation of natural and artificial Zr-silicate gels, *Scientific Basis for Nuclear Waste Management XXIX, Mater. Res. Soc. Symp. Proc.* **932**, 1017–1024.

Burakov B.E., Yagovkina M.A., Garbuzov V.M., Kitsay A.A. and Zirlin V.A. (2004) Self-irradiation of monazite ceramics: contrasting behavior of $PuPO_4$ and $(La,Pu)PO_4$ doped with Pu-238, *Scientific Basis for Nuclear Waste Management XXVIII, Mater. Res. Soc. Symp. Proc.* **824**, 219–224.

Burakov B.E., Domracheva Y.V., Zamoryanskaya M.V., Petrova M.A., Garbuzov V.M., Kitsay A.A. and Zirlin V.A. (2009) Development and synthesis of durable self-glowing crystals doped with plutonium. *J. Nucl. Mater.* **385**(1), 134–136.

Elwell D. (1979) *Man-made Gemstones* (Ellis Horwood Ltd., Publishers).

Glagovskiy E.M., Yudintsev S.V., Kouprine A.V., Pelevin L.P., Konovalov E.E., Velichkin V.I. and Myasoedov B.F. (2001) Investigation of actinide-doped matrices obtained by self-propagating high temperature synthesis, *Radiochimia* **43**(6), 557–562 (in Russian).

Hanchar J.M., Finch R.J., Hoskin P.W.O., Watson E.B., Cherniak D.J. and Mariano A.N. (2001) Rare earth elements in synthetic zircon. 1. Synthesis and rare earth element and phosphorus doping, *Am. Miner.* **86**, 667–680.

130   *Crystalline Materials for Actinide Immobilisation*

Hanchar J.M., Burakov B.E., Anderson E.B. and Zamoryanskaya M.V. (2003). Investigation of Single Crystal Zircon, (Zr,Pu)SiO$_4$, doped with ^{238}Pu, *Scientific Basis for Nuclear Waste Management XXVI, Mater. Res. Soc. Symp. Proc.* **757**, 215–225.

Hanchar J.M., Burakov B.E., Zamoryanskaya M.V., Garbuzov V.M., Kitsay A.A. and Zirlin V.A. (2004) Investigation of Pu Incorporated into Zircon Single Crystal, *Scientific Basis for Nuclear Waste Management XXVIII, Mater. Res. Soc. Symp. Proc.* **824**, 225–236.

Kitsay A.A., Garbuzov V.M. and Burakov B.E. (2004) Synthesis of actinide-doped ceramics: from laboratory experiments to industrial scale technology, *Scientific Basis for Nuclear Waste Management XXVII, Mater. Res. Soc. Symp. Proc.* **807**, 237–242.

Kulyako Y.M., Perevalov S.A., Vinokurov S.E., Myasoedov B.F., Petrov G.A., Ozhovan M.I., Dmitriev S.A. and Sobolev I.A. (2001) Properties of host matrices with incorporated U and Pu oxides, prepared by melting of a zircon-containing heterogeneous mixture (by virtue of exo effect of burning metallic fuel), *Radiochemistry* **43**(6), 626–631.

Lee W.E. and Rainforth W.M. (1994) *Ceramic Microstructures: Property Control by Processing* (Chapman and Hall, London), 604 pp.

Maddrell E.R. and Abraitis P.K. (2004) Ceramic wasteforms for the conditioning of spent MOx fuel wastes, *Scientific Basis for Nuclear Waste Management XXVII, Mater. Res. Soc. Symp. Proc.* **807**, 231–236.

Ojovan M.I. and Lee W.E. (2007) *New Developments in Glassy Nuclear Wasteforms* (Nova Science Publishers, New York).

Ojovan M.I., Petrov G.A., Stefanovsky S.V. and Nikonov B.S. (1999) Processing of large-scale radwaste-containing blocks using exothermic metallic mixtures, *Scientific Basis for Nuclear Waste Management XXII, Mater. Res. Soc. Symp. Proc.* **556**, 239–245.

Richerson D.W. (1992) *Modern Ceramic Engineering*, 2nd edn. (Marcel Dekker, New York).

Ringwood A.E., Kesson S.E., Reeve K.D., Levins D.M. and Ramm E.J. (1988) Synroc. in *Radioactive Waste Forms for the Future*, eds. W. Lutze and R. C. Enmg (North-Holland Physics Publishing, the Netherlands), pp. 233–334.

Sepelak V. (2002) Nanocrystalline materials prepared by homogeneous and heterogeneous mechanochemical reactions, *Ann. Chim. Sci. Mat.* **27**(6), 61–76.

Sobolev I.A., Stefanovsky S.V., Youdintsev S.V., Nikonov B.S., Omelianenko B.I. and Mokhov A.V. (1997) Study of melted Synroc doped with simulated high-level waste, *Scientific Basis for Nuclear Waste Management XX, Mater. Res. Soc. Symp. Proc.* **465**, 363–370.

Ushakov S.V., Burakov B.E., Garbuzov V.M., Anderson E.B., Strykanova E.E., Yagovkina M.M., Helean K.B., Guo Y.X., Ewing R.C and Lutze W. (1998) Synthesis of Ce-doped zircon by a sol-gel process, *Scientific Basis for Nuclear Waste Management XXI, Mater. Res. Soc. Symp. Proc.* **506**, 281–288.

Xu H. and Wang Y. (2000) Crystallization sequence and microstructure evolution of Synroc samples crystallized from $CaZrTi_2O_7$ melts *J. Nucl. Mater.* **279**, 100–106.

Yudintsev S.V., Ioudintseva T.S., Mokhov A.V., Nikonov B.S., Konovalov E.E., Perevalov S.A., Stefanovsky S.V., Ptashkin A.G., Glagovskiy E.M. and Kouprine A.V. (2004) Study of pyrochlore and garnet-based matrices for actinide waste produced a self-propagating high-temperature synthesis, *Scientific Basis for Nuclear Waste Management XXVII, Mater. Res. Soc. Symp. Proc.*, **807**, 273–278.

Chapter 5

Examination of Highly Radioactive Samples

5.1. XRD Analysis

Bulk powder X-ray diffraction analysis (XRD) is commonly used to identify the crystal phase composition of solids (Cullity, 1956). Aspects of XRD of highly radioactive actinide-bearing samples that need to be emphasised include:

(i) Preparation of samples for XRD usually involves grinding with or without an internal standard (e.g., quartz, diamond, crystalline polymorphs of Al_2O_3) in an agate mortar to give a fine homogeneous powder. However, handling of actinide-bearing powders containing, for example, ^{238}Pu, ^{244}Cm and ^{241}Am is difficult even in specially designed glove boxes. To avoid sample grinding, in some cases thin ceramic pellets instead of powder are used;

(ii) Operation of a shielded XRD unit in a hot cell or large-scale glove box is extremely expensive. Regular unshielded XRD equipment for analyses of hazardous samples may be used if the samples are sealed in a special hermetic holder or cassette. Such samples might be considered sealed sources of radiation and handled outside a glove box. PNNL has developed and applied such a technique to XRD of ^{238}Pu-doped ceramics (Strachan *et al.* 2004, 2005, 2008). KRI has developed a sample-holding cassette with a metallic beryllium window, which is transparent to X-rays, as

134   *Crystalline Materials for Actinide Immobilisation*

Fig. 5.1.1.   Parts of cassette developed by KRI (Burakov, 2000) for XRD analysis of highly radioactive actinide-doped samples: regular sample-holder (left) for loading powdered sample and internal standard or thin solid ceramic pellets; and aluminium lid (right) with circular metallic beryllium window. Radioactive sample is loaded into cassette and hermetically sealed with the lid in a glove box. After surface decontamination the cassette is considered to be an SRS.

shown in Fig. 5.1.1 (Burakov, 2000). The cassette is sealed by welding or using temporary glue in a glove box; use of glue is preferable if the sample is to be removed for further study after XRD analysis. After cleaning and checking the surface for contamination, the sample can be regarded as a Sealed Radioactive Source (SRS, as used in hospitals and research). The use of such cassettes has allowed KRI to perform repeated XRD analysis of the same ^{238}Pu-doped ceramic samples with increasing cumulative dose over 5–6 years (Burakov *et al.*, 2008).

5.2. SEM and EPMA

Similar procedures for scanning electron microscopy (SEM) and electron probe microanalysis (EPMA) of actinide-bearing samples are performed as for non-radioactive samples, as described in standard texts (e.g., Goldstein *et al.*, 1981), although with some differences.

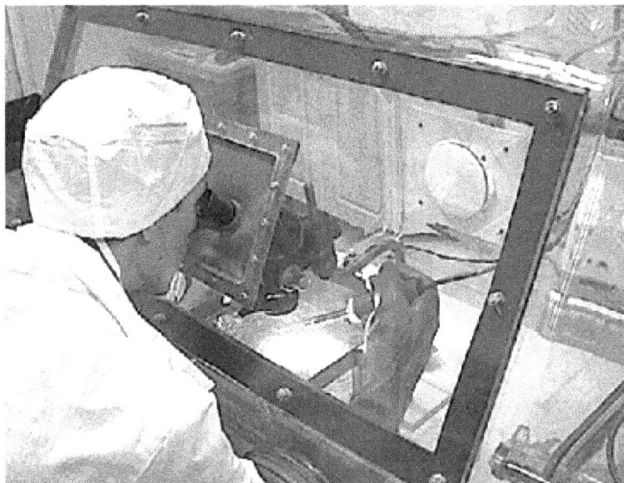

Fig. 5.2.1.   Preliminary examination and selection of "hot" Chernobyl particles for further EPMA and SEM using an optical microscope installed into a glove-box (Burakov *et al.*, 2003).

For example, equipment (in particular the vacuum camera) should be protected from radiation damage and contamination. Consequently the sample size is minimised to limit the amount of active material utilised and to limit release of dust particles and aerosols. Representative samples for SEM and EPMA should be selected using an optical microscope (Fig. 5.2.1).

Selected ceramic sections or crystals for quantitative EPMA are mounted in epoxy or acrylic resin to prepare a "pellet" with polished surface (Fig. 5.2.1). Epoxy is optically transparent and therefore the orientation of samples can be checked and corrected (if necessary) before it solidifies. Acrylic resin (as used by dentists for filling teeth) is not transparent and solidifies rapidly. It is also more brittle, mechanically durable and resistant to alpha-irradiation compared to an epoxy matrix. Radioactive pellets are usually polished manually (under glove-box conditions) and this procedure is easier for acrylic resin pellets.

Polished pellets are cleaned and checked for surface contamination using cotton cleaning smears (to check for alpha-radiation). If necessary, this procedure may be repeated several times. If there is

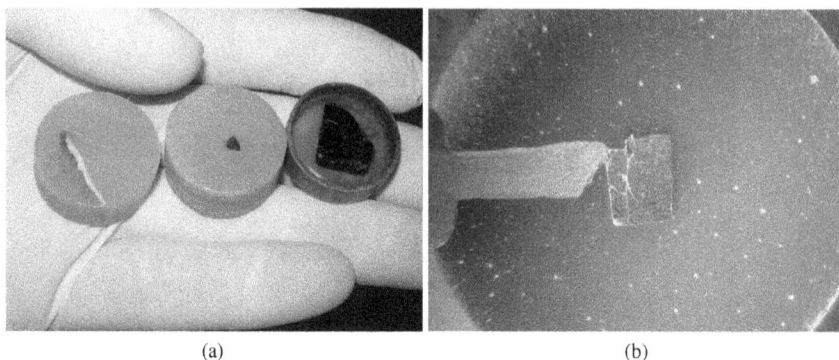

(a)   (b)

Fig. 5.2.2.   Uncoated sample-holders or pellets used for complex analyses by optical microscopy, EPMA and cathodoluminescence imaging and spectroscopy: general image of two pellets (left and centre in (a) with acrylic resin matrix and in epoxy (right in (a)) and closer image of polished pellet surface with sample in central part and strip of electroconductive varnish (b).

any doubt about the quality of decontamination, it is possible to paint unpolished areas of the pellet surface with electroconductive varnish.

Coating with an electrical conductor (such as C or Au) is not required if the sample matrix is electrically conducting (for example, UO_2). In this case a strip of electroconductive varnish between sample and metallic support is sufficient to enable SEM imaging and qualitative EPMA (Fig. 5.2.2) without any charging problem (Lee and Rainforth, 1994).

Prior to any coating the "pellet" should be examined in an optical microscope in reflected light (Fig. 5.2.3). All inhomogeneous areas in matrices of ceramics or crystals have to be recorded and the orientation of single crystals in the matrix of polished pellets documented. Regions chosen for quantitative EPMA should be away from cracked zones but they do have to be representative of the variation of chemical composition in the whole sample matrix (Fig. 5.2.4 and Table 5.2.1).

Intense alpha radiation may destroy the deposited electroconductive film and make SEM imaging and EPMA impossible. In some cases radiation from actinide-bearing samples affects the semiconductor signal detectors (e.g., secondary or backscattered electrons),

(a)

(b)

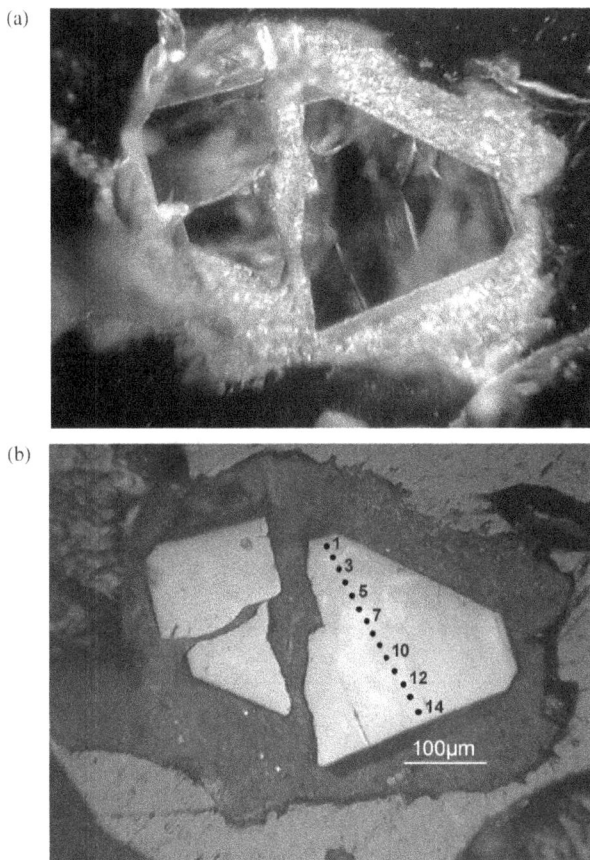

Fig. 5.2.3.   Reflected light images of three pieces of the same [238]Pu-doped zircon single crystal mounted in acrylic resin. Taken in dark field, (a) clearly demonstrates the crystal orientation despite its cracking into separate pieces as a result of radiation damage. The grain in (b) was marked by points in the EPMA, which avoided the cracks, analysed in Fig. 5.2.4. Analysis from these points is given in Fig. 5.2.4.

which leads to poor image quality. To overcome this problem SEM and EPMA may be done immediately after coating or coating thickness can be increased and SEM imaging done at higher accelerating voltage and beam current. Optimal film thickness should be determined experimentally taking into account the fact that too thick a coating degrades image resolution. For precise quantitative EPMA,

138   *Crystalline Materials for Actinide Immobilisation*

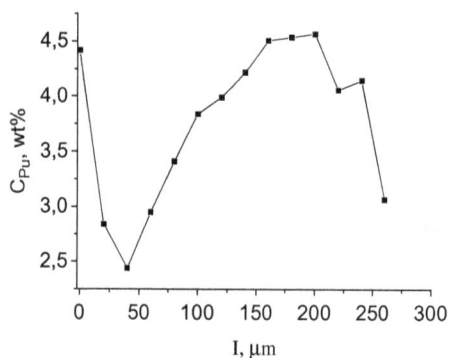

Fig. 5.2.4. EPMA profile (from 1st to 14th point) through a cross-section of ^{238}Pu-doped zircon crystal (see Fig. 5.2.3 and Table 5.2.1). C_{Pu} is Pu concentration and I distance from first point.

Table 5.2.1. Results of EPMA (in wt.% el.) of cross-section of ^{238}Pu-doped zircon crystal (see Figs. 5.2.3 and 5.2.4).

Point	Si	Zr	Pu
1	15.3	47.7	4.4
2	15.1	48.1	2.8
3	14.9	48.1	2.4
4	15.0	47.7	3.0
5	15.0	47.6	3.4
6	14.9	46.8	3.8
7	14.8	46.9	4.0
8	15.1	46.2	4.2
9	14.9	45.9	4.5
10	14.9	45.8	4.5
11	14.7	45.8	4.6
12	14.9	46.6	4.1
13	15.4	46.3	4.2
14	15.3	47.7	3.1

separate pellets with EPMA standards and samples should be coated simultaneously. This method was used by KRI for examination of Chernobyl "lava" and "hot" particles (Burakov, 1993; Burakov *et al.*, 1994 and Burakov *et al.*, 2003). Coating with a thick copper film provided protection against irradiation of the semiconductor detectors.

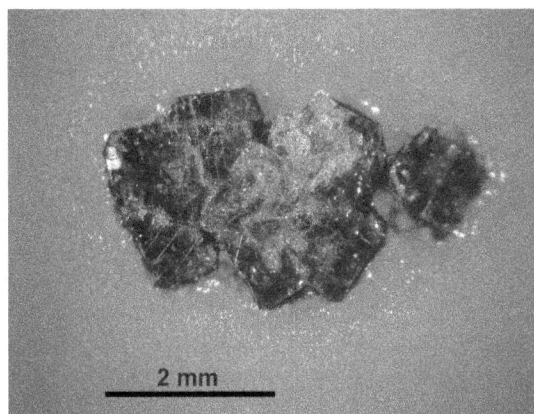

Fig. 5.2.5.   Reflected light image of aggregate of ^{238}Pu-doped zircon crystals mounted in acrylic resin. Six years after its preparation, intensive self-irradiation caused sample swelling and cracking. The acrylic resin matrix was also affected by irradiation causing formation of a damaged ring around the zircon sample.

At the same time this allowed high-contrast backscattered electron images to be obtained. A similar technique was applied by KRI to carry out quantitative EPMA of Ti-pyrochlore and cubic zirconia ceramics doped with 8–10 wt.% ^{238}Pu (Burakov *et al.*, 2002; Zamoryanskaya and Burakov, 2004).

Damage to the acrylic or epoxy resins from alpha-radiation (Fig. 5.2.5) should be taken into consideration in samples that require repeated SEM and EPMA after accumulation of increasing dose.

Single crystals of PuO_2, NpO_2 (Figs. 1.3.4. and 1.3.5), etc., are used as standards for quantitative actinide EPMA.

5.3. Cathodoluminescence

When some solids (e.g., insulators and semiconductors) are bombarded with electrons, long-wavelength photons are emitted in the UV and visible regions of the electromagnetic spectrum, a phenomenon known as cathodoluminescence (CL). Many modern SEMs and EPMAs have detectors for this light and can operate in CL mode. Some of these instruments are equipped with an optical microscope which allows direct visual CL observation on areas from a few to

140   *Crystalline Materials for Actinide Immobilisation*

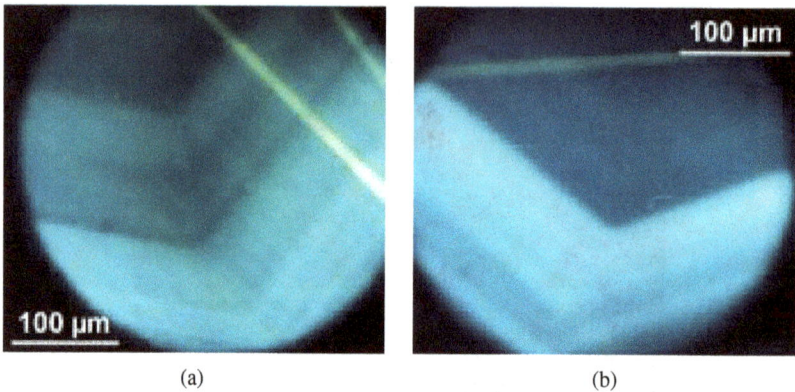

Fig. 5.3.1. Cathodoluminescence images of zoned areas in crystals of ^{239}Pu-doped zircon (Burakov *et al.*, 2002). Plutonium content ranged from 0.1 wt.% (in light zones) to 1.4 wt.% (dark zones). This zoned inhomogeneity of zircon crystals is not observed by other methods such as optical microscopy and backscattered electron SEM imaging.

hundreds of microns. The sample may be studied by SEM, EPMA and CL methods using the same multi-detector equipment. CL images can be recorded by a camera installed into an optical port of an SEM and EPMA. The same optical port might be used to install a compact CL spectroscope (Zamoryanskaya *et al.*, 2004), which allows CL-spectra to be obtained from (1–3 μm) regions of the sample.

CL imaging can be more sensitive to some chemical inhomogeneities' defects compared to backscattered electron SEM imaging and optical microscopy (Fig. 5.3.1). CL spectroscopy is an important tool in determining the valence states of actinides incorporated into the crystalline structure of host-phases (Zamoryanskaya and Burakov, 2006) or dissolved into glass matrices. CL spectra of anisotropic crystalline phases are a function of crystal orientation (Burakov *et al.*, 2007). CL can be used to determine crystal orientation in actinide-bearing single crystals which could not be obtained on pellet preparation.

Depending on cumulative dosage, CL emission in actinide host-phases is affected by radiation damage. In the simplest cases, the intensity of CL emission is a function of dose and loss of the crystallinity (Fig. 5.3.2). Surprising results have been obtained on

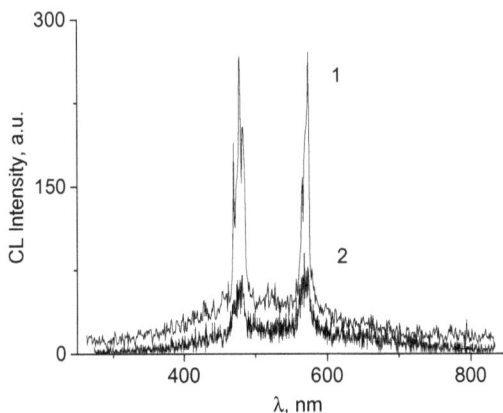

Fig. 5.3.2.   Evolution of cathodoluminescence spectra of zircon doped with 2.4–2.7 wt.% ^{238}Pu (see Fig. 2.3.4(a)): shortly after zircon synthesis (top spectrum) and after 574 days (lower spectrum). Courtesy of Dr. M.V. Zamoryanskaya of the Ioffe Physical-Technical Institute, St. Petersburg, Russia.

repeated CL study of Ti-pyrochlore ceramics doped with ^{238}Pu and ^{238}U (Zamoryanskaya *et al.*, 2002). Self-irradiation of the pyrochlore phase, $(Ca,Gd,Hf,Pu,U)_2Ti_2O_7$, was accompanied by formation and accumulation of uranyl-ion, $(UO_2)^{2+}$, and tetrahedral U-complexes, $(UO_4)^{2-}$. CL spectroscopy can thus reveal the evolution of actinide speciation, a phenomenon not seen by other techniques.

Additional application of the CL method has been suggested for development of self-glowing crystals doped with small amounts of actinides (Burakov *et al.*, 2007; 2009). The intensity of CL emission initiated by non-radioactive "phosphor-elements" (Eu,In,Tb,Dy, etc.) incorporated into the structures of, for example, zircon, garnet, xenotime and monazite crystals correlates with the self-glowing intensity of the same host-phases doped with the same phosphors and also actinides. This observation suggests the optimum level of phosphor admixture might be identified from a CL study of non-radioactive crystals before experiments on active crystals. Early results on application of CL in this field are summarised in Table 2.4.1. (see Chap. 2). They confirm the optimum amount of phosphor generates intense self-glowing while decreasing the ^{238}Pu content needed.

5.4. Optical Microscopy

Optical microscopy of ceramics is well described in books (for example, Bousfield, (1992)) and reviews such as Roberts and Robinson, (1955), although it is not applied to the study of actinide hosts as much as it should be. Some important features of crystalline materials doped with actinides are easily observed using a simple optical microscope (Fig. 5.4.1). For example, changes in colour and crack formation have been detected in polycrystalline ceramics and single crystals doped with ^{238}Pu (Burakov *et al.*, 2009). For example, the colour of a monazite ceramic, $(La,Pu)PO_4$, became grey

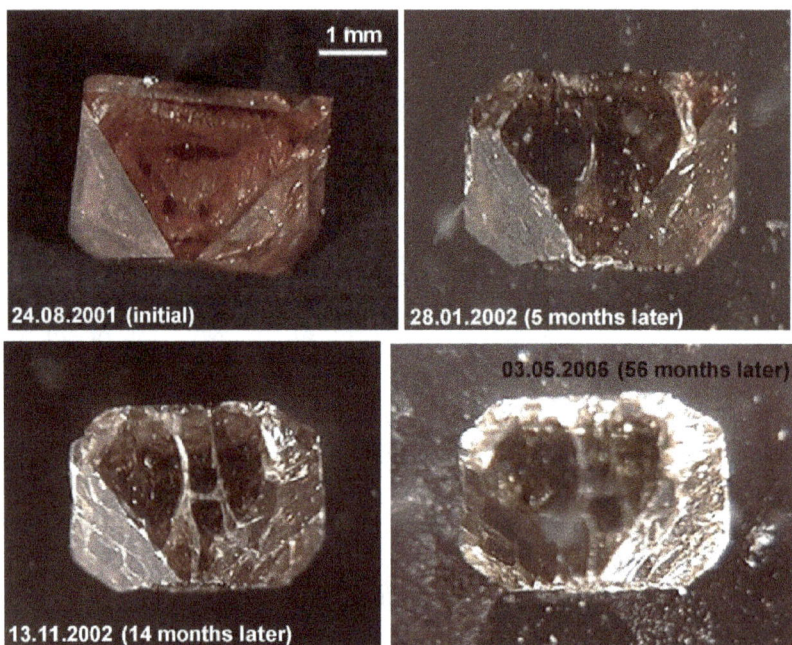

Fig. 5.4.1.   Behaviour of ^{238}Pu-doped zircon single crystal under self-irradiation: this crystal was placed in a hermetic glass cassette for repeated optical microscope study. Average bulk content of ^{238}Pu is 2.4 wt.%, but the distribution of Pu (measured by EPMA for all Pu isotopes) is inhomogeneous and ranges from 1.9 to 4.7 wt.%. The lowest Pu concentrations occur at the outer edges of the crystal. Self-irradiation caused crystal cracking and change of colour from initial pink-brown to brown, grey-brown, yellow-grey and green-grey. Formation of small (from few to dozens μm) particles around the crystal was clearly observed (Burakov *et al.*, 2009).

from initial light blue under self-irradiation from ^{238}Pu although it remained crystalline (from XRD) and did not crack. Polycrystalline Pu-monazite, $PuPO_4$, underwent metamictisation, cracked into separate pieces and changed colour from initial deep blue to black. Zircon/zirconia ceramic did not crack under self-irradiation and preserved its initial light grey colour despite becoming completely amorphous. However, ^{238}Pu-doped zircon single crystal (Figs. 5.4.1 and 5.4.2) examined over a 7-year period developed matrix cracking and changed from initial pink-brown to brown, grey-brown, yellow-grey and then green-grey. Another important feature of this crystal was its self-glowing in the dark (Fig. 2.4.1), which eventually disappeared due to the radiation damage. All actinide-doped samples at KRI, which require repeated investigation by different methods over long times, are periodically examined by optical microscopy (Fig. 5.4.3). In some cases investigation may need to be halted if the sample has disintegrated so much that it becomes a source of radioactive particles which may contaminate equipment (Fig. 5.2.5).

5.5. Mechanical Durability

Characterisation of mechanical stability, which is so important for non-radioactive ceramics (Wachtman, 1996), is not typically done for radioactive wasteforms including highly radioactive actinide-doped crystalline materials (Ojovan and Lee, 2005). Detailed information about the mechanical integrity is available only for yttria-stabilised cubic zirconia ceramic doped with 20, 30 and 40 mol.% NpO_2 (Kinoshita *et al.*, 2006). Predominantly single-phase ceramics with good mechanical and chemical stability include alumina and zirconia as a result of which they are used, for example, as prostheses in the human body. The use of multi-phase ceramic compositions typically leads to a decrease in durability as glassy grain boundary phases, for example, are less stable in aqueous environments. Incorporation of actinides into the crystal structures of host-phases may also affect their mechanical properties. It is appropriate to apply a simple and rapid technique to compare the mechanical stability of actinide-doped ceramics. A possible method to characterise a rough

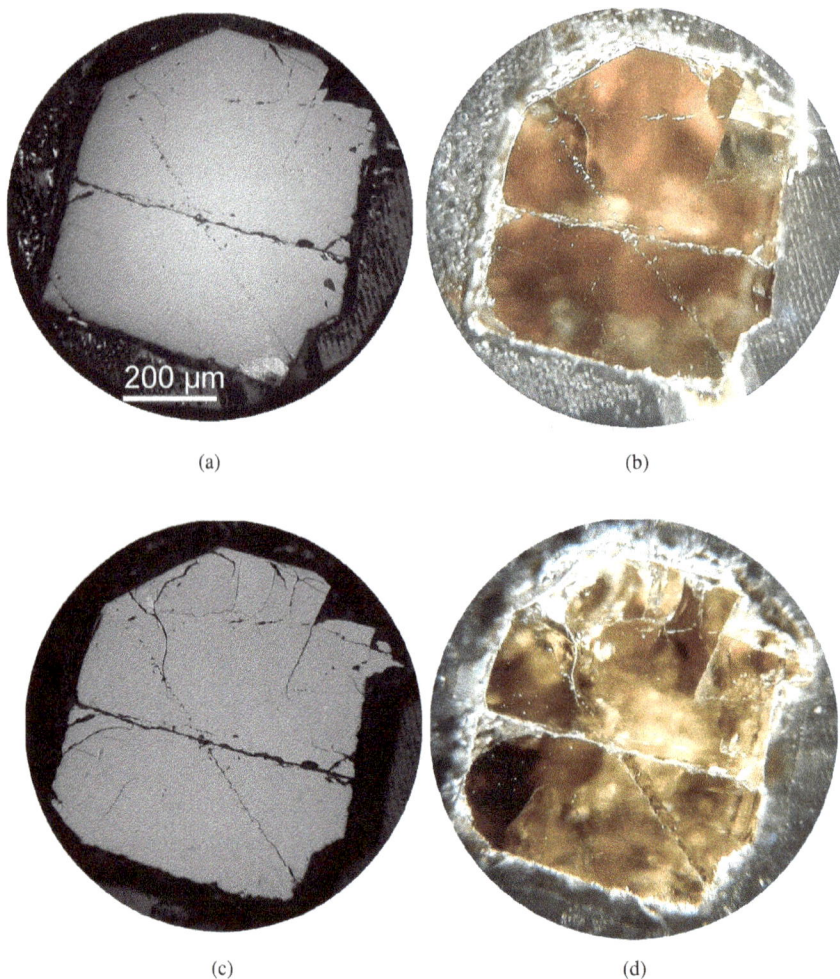

Fig. 5.4.2.   Reflected light images of zircon single crystal doped with 2.4 wt.% ^{238}Pu: shortly after zircon synthesis ((a) and (b)) and after 574 days ((c) and (d)). Images (b) and (d) have been taken in dark field to see the internal crystal matrix features. Self-irradiation caused cracking and change of colour (images (c) and (d)).

measure of mechanical stability is similar to the cold crushing (compressive) strength test historically used in the refractories industry (Chesters, 1983), but with little physical validity (as shown in Fig. 5.5.1), although it is easily performed in a glove box. Radioactive

Fig. 5.4.3.   Hermetically sealed glass-cassette developed at KRI for repeated periodic examination of highly radioactive actinide-doped samples using an optical microscope.

ceramics should be compared with well-characterised non-radioactive standards or other actinide-doped samples.

5.6. Leach and Alteration Tests

Thorough characterisation and understanding of the chemical durability of actinide-doped ceramics is a key need in the development of ceramic wasteforms. However, the current state of knowledge of materials' durability over millennia is limited and uncertainties (particularly for cement and ceramic wasteforms) must be addressed via future (international) research programmes.

A crucial measure of the reliability of radionuclide immobilisation is the rate at which radionuclides are released from the wasteform during long-term storage or permanent disposal. As the most plausible path for reintroduction of radioactivity into the biosphere is via water, the most important parameters that characterise the ability of a wasteform to hold onto the active species are the leach rates. The leaching behaviour of wasteforms containing different amounts of radionuclides is compared using the normalised leaching rates NR_i

146   *Crystalline Materials for Actinide Immobilisation*

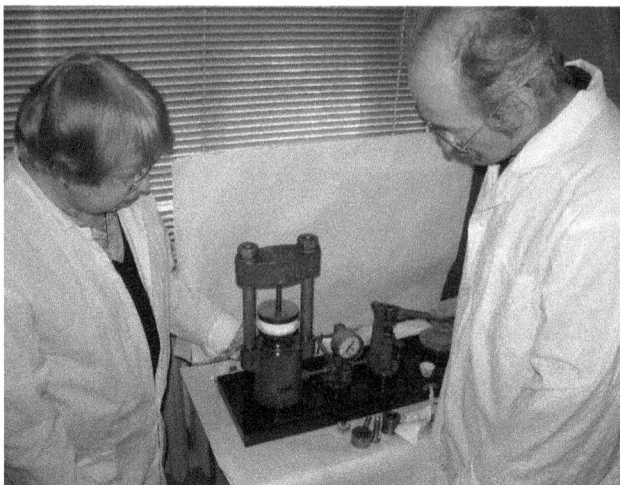

Fig. 5.5.1.   A simple check of mechanical stability of an inactive large ($\sim$10-cm diameter) ceramic pellet using hydrostatic press: ceramic stability is calculated from the ratio of pressure required to break the pellet and the ceramic pellet top surface area.

for each ith nuclide. The normalised leaching rate NR_i (in g/cm^2 day) for the ith component (radionuclide) is calculated from:

$$NR_i = \frac{A_i}{A_0} \cdot \frac{W_0}{S \cdot t} \tag{5.1}$$

where A_i is the amount of component i (in g) leached from the sample in time t (in days), A_0 is the original amount of this component in the sample (in g), W_0 is the original weight of sample (in g), S is the water contacting surface area of the sample (in cm^2). NR_i are determined using a range of approved tests such as the IAEA test protocol ISO 6961-1982. A set of standard tests to determine the water durability of vitrified waste and other wasteforms was developed at the Materials Characterisation Centre (MCC) of Pacific Northwest National Laboratory, USA. These MCC tests are now used worldwide as the internationally approved standards. The most important tests are given in Table 5.6.1 (Strachan, 2001).

It is relatively easy to confirm the low solubility of actinide host-phases using the static leach test MCC-1 (ASTM C1220-92, 1995)

Table 5.6.1.   Standard leach and alteration tests.

Test designation	Conditions	Use
ISO 6961, MCC-1	Deionised water. Static. Monolithic specimen. Sample surface to water volume (S/V) usually $10\,m^{-1}$. Open to atmosphere. Temperature 25°C (IAEA), 40, 70 and 90°C (MCC-1)	For comparison of wasteforms.
MCC-2	Deionised water. Temperature 90°C. Closed.	Same as MCC-1, but at higher temperature.
PCT (MCC-3)	Product consistency test. Deionised water stirred with glass or ceramic powder. Various temperatures. Closed.	For durable wasteforms to accelerate leaching.
SPFT (MCC-4)	Single pass flow through test. Deionised water. Open to atmosphere.	This is the most informative test.
VHT	Vapour phase hydration. Monolithic specimen. Closed. High temperatures.	Accelerates alteration product formation.

and to compare the results obtained to those from glassy wasteforms (Lutz and Ewing, 1988). MCC-1 procedures may slightly vary between different laboratories. Only a brief description of the MCC-1 modified procedure used at KRI for routine leaching experiments, performed outside a glove-box environment, is given here. Ceramic samples of known composition and surface area are placed on the bottom of a Teflon^{TM} test vessel (Fig. 5.6.1(a)) with deionised or distilled water and then held at fixed temperature in an oven. MCC-1 leach tests are normally performed at 25 and 90°C. Leaching at such temperatures does not require special protection against pressurised water that may be heavily contaminated by actinides. In some cases it is useful to go above 100°C and use acids to speed up the leaching-alteration processes. These experiments are carried out in Teflon^{TM} vessels placed in a stainless steel container (Fig. 5.6.1(b)) or autoclave (Fig. 5.6.2).

148   *Crystalline Materials for Actinide Immobilisation*

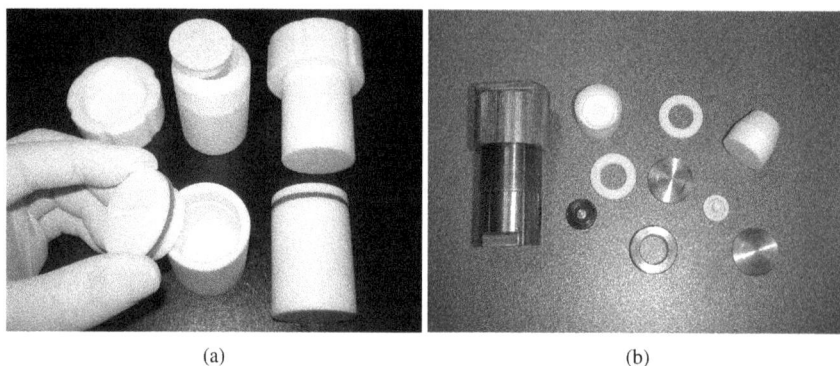

(a)   (b)

Fig. 5.6.1.   (a) Vessels used for static leach test of actinide-doped ceramics: two types of Teflon^{TM} vessel for experiments at a temperature up to 90°C and (b) stainless steel pressure vessel with internal Teflon^{TM} core made of two parts for experiments at ≤120–150°C.

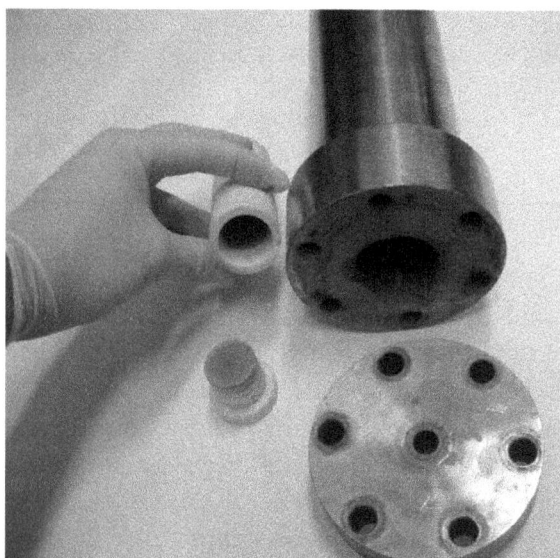

Fig. 5.6.2.   Stainless steel autoclave with internal Teflon^{TM} vessel for alteration experiments at 150–190°C.

To avoid direct contact between ceramic and Teflon^{TM} the pellet is fixed by Pt-wire or placed on Pt-foil (Fig. 5.6.3).

The ratio of sample surface area to water volume is maintained at 1:10 and the usual test duration is 28 days. After 3 and 14 days, the

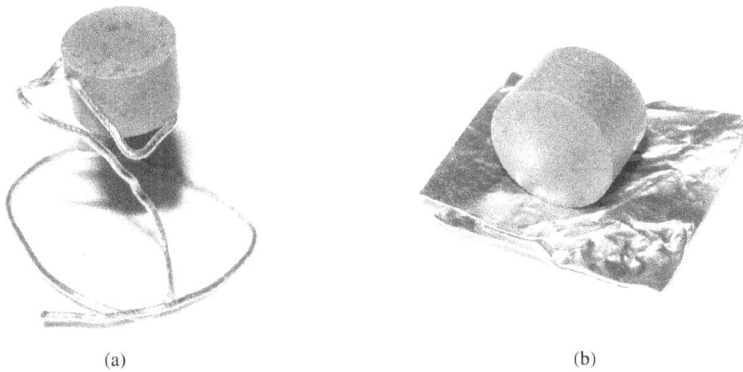

(a)   (b)

Fig. 5.6.3. Two possible arrangements of ceramic pellet at the bottom of Teflon^{TM} vessel for static leach test: on (a) Pt-wire or (b) Pt-foil.

leaching solution is removed and replaced by fresh deionised water. The specimens are rinsed with high purity water and ethanol. Analysis of actinides in the leaching solution is carried out after 3, 14 and 28 days using spectrometric equipment (Sobolev and Belyaev, 2002). The actinides absorbed on the walls of vessels are extracted using concentrated acid and measured after 28 days, then the correction to the activity of Pu and Am released by the 3rd and 14th day is provided, assuming a linear relation of the absorbed amount with time for the first 14 days. Because of the typically very low ceramic solubility level it is possible to carry out leach tests without additional sampling after 3 and 14 days. Normalised Pu mass loss (NL), a relative measure of leach rate, is calculated from:

$$NL = \frac{A}{A_0} \frac{W_0}{S},\tag{5.2}$$

where A is the total activity (Bq) of actinide element in the water solution after leaching, A_0 is the initial activity (Bq) of actinide element in the specimen, W_0 is the initial mass (in g) of the specimen, S is the sample surface area (m^2) based on external dimension without correction for porosity. Some data from MCC-1 leach tests are given in Table 5.6.2.

The static leach test enables early identification of ceramics containing separate actinide phases with high leach rates (Burakov *et al.*, 2001, and Nikolaeva and Burakov, 2002). However, the MCC-1 test

150   *Crystalline Materials for Actinide Immobilisation*

Table 5.6.2.   Normalised Pu mass losses, NL(Pu), from matrices of ^{238}Pu-doped ceramics based on: gadolinia-stabilised cubic zirconia (9.9 wt.% ^{238}Pu); zircon (4.6 wt.% ^{238}Pu) and Ti-pyrochlore (8.7 wt.% ^{238}Pu) after leach testing (in deionised water at 90°C for 28 days) depending on cumulative dose (Burakov *et al.*, 2009).

Cumulative dose in ceramic doped with ^{238}Pu, alpha- decays/m$^3 \times 10^{23}$	NL(Pu), g/m^2	Equivalent years of storage calculated for the ceramic doped with ^{239}Pu
Cubic zirconia ceramic with density 5.6 g/cm^3 (96% of theoretical density)		
11	0.04	30
56	0.35	140
81	0.37	200
127	0.24	320
Zircon-based ceramic with density 4.4 g/cm^3 (93% of theoretical density)		
7	0.01	30
31	0.04	150
43	0.05	210
66	0.04	330
Ti-pyrochlore based ceramic with density 4.9 g/cm^3 (82% of theoretical density)		
29	0.22	80
49	0.28	140
100	0.84	280
133	1.93	380

does not help in the selection of the most durable ceramics because many different samples have similar, but low actinide release. For example, dozens of static leach tests (28 days at 90°C in deionised water) using samples of ^{239}Pu-doped (5–10 wt.% Pu) ceramics based on Ti-pyrochlore $(Ca,Gd,Hf,Pu,U)_2Ti_2O_7$, zircon/zirconia $(Zr,Pu)SiO_4/(Zr,Pu)O_2$, and cubic zirconia $(Zr,Gd,Pu)O_2$, gave similar NL(Pu) $= 10^{-4}$–10^{-3} g/m^2. These data are similar to results from static leach tests for ^{239}Pu-doped multi-phase titanate ceramics based on zirconolite and pyrochlore (Hart *et al.*, 2000) and single-phase thorium phosphate-diphosphate (TPD) ceramic (Dacheux *et al.*, 1999). NL of ^{241}Am, ^{238}Pu and ^{238}U from ceramics under the same conditions is in general at least 10 times higher than for ^{239}Pu. This was the case for NL(Am) even in zircon/zirconia, cubic zirconia and Ti-pyrochlore ceramics doped with only 0.1 wt.% ^{241}Am.

For NL(^{238}Pu) it is true even for low cumulative self-irradiation dose (Table 5.6.2). In contrast, standard leach tests (28 days at 90°C in deionised water) with non-radioactive actinide surrogates such as Ce, Eu, Gd demonstrate the usual NL $= 10^{-5}$–10^{-4} g/m^2.

Clearly, leach tests performed in distilled and deionised water do not give an understanding of the behaviour of ceramic wasteforms under conditions of geological disposal. Underground water which is an aqueous solution saturated with different ions from the host rock may contact actinide-bearing wasteforms at depths of 300 m to 1–2 km. To model the interaction between ion-saturated solutions and actinide-doped ceramics requires use of a powdered sample of actinide-doped ceramic (Burakov and Anderson, 2002), which interacts (at 90°C and higher) with an aqueous solution simulating underground water of the prospective repository site (Fig. 5.6.4). Experiments last for 3–6 months. The phase composition of altered ceramic powder is characterised by XRD. Bulk ground rock samples and single-mineral grains are studied for actinide sorption. New phases in the rock sample and the actinide content of the solution might also be investigated. Similar experiments could be carried out using a vessel made of monolithic rock (granite) and a solid pellet

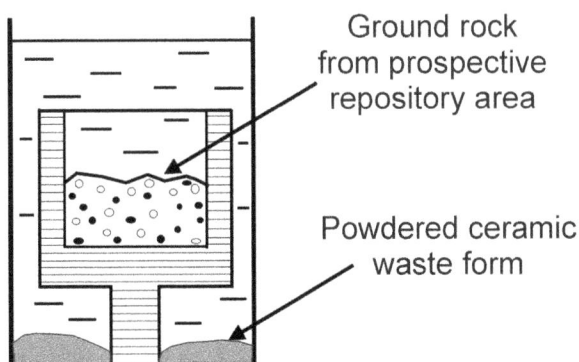

Fig. 5.6.4.   Schematic cross-section of vessel used for alteration-sorption experiment in ion-saturated solution simulating underground water (Burakov and Anderson, 2002). After the experiment, bulk samples of altered rock and actinide-doped ceramic are studied by XRD to identify newly formed phases. Single mineral phases in altered rock may be studied separately for actinide sorption.

152   *Crystalline Materials for Actinide Immobilisation*

Fig. 5.6.5.   Special vessel made of granite core for actinide sorption-diffusion experiment. A pellet of actinide-doped ceramic is placed at the bottom of this vessel, which is then filled with distilled water and hermetically sealed. Experiments last for several months at $\geq 90°$C.

instead of powder samples (Fig. 5.6.5). In this case the actinide diffusion-sorption process, which takes place in the host rock matrix, might be studied. After the experiment the granite vessel is sectioned with a diamond blade and both parts are studied by autoradiography and optical microscopy. Redistribution of actinides on the surface of a particular mineral phase of the host rock might be identified.

References

ASTM C1220-92 (1995) Standard Test Method for Static Leaching of Monolithic Waste Forms for Disposal of Radioactive Waste. *Annual Book of ASTM Standards, Philadelphia: American Society for Testing and Materials*, 710–724.

Bousfield B. (1992) *Surface Preparation and Microscopy of Materials*. Wiley, New York.

Burakov B. (1993) A study of high-uranium technogeneous zircon $(Zr,U)SiO_4$ from Chernobyl 'lavas' in connection with the problem of creating a crystalline matrix for high-level waste disposal, *Proc. Intern. Conf. SAFE WASTE'93*, 13-18/06/1993, Avignon, France, **2**, 19–28.

Burakov B.E., Anderson E.B., Galkin B.Ya., Pazukhin E.M. and Shabalev S.I. (1994) Study of Chernobyl "hot" particles and fuel containing masses: implications for reconstruction of the initial phase of the accident, *Radiochimica Acta* **65**, 199–202.

Burakov B.E. (2000) KRI Studies of the US Pu Ceramics [B506203]. in Jardine, L.J. and Borisov G.B. (eds). *Excess Weapons Plutonium Immobilization in Russia*, pp. 251–252, Proc. Meet. for Coordination and Review of Work, St. Petersburg, Russia, 1999, UCRL-ID-138361.

Burakov B.E., Anderson E.B., Zamoryanskaya M.V., Yagovkina M.A., Strykanova E.E. and Nikolaeva E.V. (2001) Synthesis and study of ^{239}Pu-doped ceramics based on zircon, (Zr,Pu)SiO$_4$, and hafnon, (Hf,Pu)SiO$_4$. in *Scientific Basis for Nuclear Waste Management XXIV, Mater. Res. Society Symp. Proc.* **663**, 307–313.

Burakov B.E., Hanchar J.M., Zamoryanskaya M.V., Garbuzov V.M. and Zirlin V.A. (2002) Synthesis and investigation of Pu-doped single crystal zircon, (Zr,Pu)SiO$_4$. *Radiochimica Acta* **89**, 1–3.

Burakov B.E., Anderson E.B., Yagovkina M., Zamoryanskaya M. and Nikolaeva E. (2002) Behavior of ^{238}Pu-doped ceramics based on cubic zirconia and pyrochlore under radiation damage. *J. Nucl. Sci. and Technology*, Suppl. 3, 733–736.

Burakov B.E. and Anderson E.B. (2002) Durability of actinide ceramic waste forms under conditions of granitoid rocks. *CD-ROM Proc. Intern. Conf. Waste Management '02*, Tucson, Arizona, USA.

Burakov B.E., Shabalev S.I. and Anderson E.B. (2003) Principal features of Chernobyl hot particles: phase, chemical and radionuclide compositions. in Barany S. (ed). *Role of Interfaces in Environmental Protection* Vol. 24, pp. 145–151, Kluwer Academic Publishers, NATO Science Series, Earth and Environmental Sciences.

Burakov B.E., Garbuzov V.M., Kitsay A.A., Zirlin V.A., Petrova M.A., Domracheva Ya.V., Zamoryanskaya M.V., Kolesnikova E.V., Yagovkina M.A. and Orlova M.P. (2007) The use of cathodoluminescence for the development of durable self-glowing crystals based on solid solutions YPO$_4$-EuPO$_4$, *Semiconductors* **41**(4), 427–430.

Burakov B.E., Yagovkina M.A., Zamoryanskaya M.V., Garbuzov V.M., Zirlin V.A. and Kitsay A.A. (2008) Self-irradiation of ceramics and single crystals doped with Pu-238: summary of 5 years of research of the V.G. Khlopin Radium Institute. in *Scientific Basis for Nuclear Waste Management, Mater. Res. Society Symp. Proc.* **1107**, 381–388.

Burakov B.E., Domracheva Ya.V., Zamoryanskaya M.V., Petrova M.A., Garbuzov V.M., Kitsay A.A. and Zirlin V.A. (2009) Development and synthesis of durable self-glowing crystals doped with plutonium. *J. Nucl. Mater.* **385**(1), 134–136.

Chesters J.H. (1983) *Refractories Production and Properties*. Materials Society, London.

Cullity B.D. (1956) *Elements of X-ray Diffraction*, Addison Wesley, New York.

154   *Crystalline Materials for Actinide Immobilisation*

Dacheux N., Thomas A.C., Chassigneux B., Pichot B., Brandel V. and Genet M. (1999) Study of $Th_4(PO_4)P_2O_7$ and solid solutions with U(IV), Np(IV) and Pu(IV): synthesis, characterization, sintering and leaching tests. in *Proc. Scientific Basis for Nuclear Waste Management XXII, Mater. Res. Society Symp. Proc.* **556**, 85–92.

Goldstein J.I., Newbury D.E., Echlin P., Joy D.C., Fiori C. and Lifshitz E. (1981) *Scanning Electron Microscopy and X-Ray Microanalysis: A Text for Biologists, Materials Scientists, and Geologists*, Plenum Press. New York.

Hart K.P., Zhang Y., Loi E., Aly Z., Stewart M.W., Brownscombre A., Ebbinghaus B.B. and Bourcier W. (2000) Aqueous durability of titanate ceramics designed to immobilise excess plutonium. in *Proc. Scientific Basis for Nuclear Waste Management XXIII, Mater. Res. Society Symp. Proc.* **608**, 353–358.

Kinoshita H., Kuramoto K., Uno M., Yamanaka S., Mitamura H. and Banba T. (2006) Mechanical integrity of yttria-stabilized zirconia doped with Np oxide. in *Scientific Basis for Nuclear Waste Management XXIX, Mater. Res. Society Symp. Proc.* **932**, 647–654.

Lee W.E. and Rainforth W.M. (1994) *Ceramic Microstructures: Property Control by Processing*, pp. 604, Chapman and Hall, London.

Lutze W. and Ewing R.C. (1988) *Radioactive Waste Forms for the Future*, North-Holland Physics Publishing, the Netherlands.

Nikolaeva E.V. and Burakov B.E. (2002) Investigation of Pu-doped ceramics using MCC-1 leach test. in *Scientific Basis for Nuclear Waste Management XXV, Mater. Res. Society Symp. Proc.* **713**, 429–432.

Ojovan M.I. and Lee W.E. (2005) *An Introduction to Nuclear Waste Immobilisation*, pp. 315, Elsevier Science Publishers, Amsterdam.

Roberts E.W. and Robinson P.C. (1955) Light microscopy of ceramics. *J. Microscopy* **140**, 137–158.

Sobolev I.A. and Belyaev E.N. (eds). (2002) *Guidelines on Environmental Radioactivity Control*, pp. 432, Meditzina, Moscow.

Strachan D.M. (2001) Glass dissolution: testing and modelling for long-term behaviour. *J. Nucl. Mat.* **298**, 69–77.

Strachan D.M., Scheele D.M., Icenhower J.P., Buck E.C., Kozelisky A.E., Sell R.L., Elovich R.J. and Buchmiller W.C. (2004) Radiation damage effects in candidate ceramics for plutonium immobilization: Final Report. *PNNL-14588*, Richland, Washington 99352.

Strachan D.M., Scheele R.D., Buck E.C., Icenhower J.P., Kozelisky A.E., Sell R.L., Elovich R.J. and Buchmiller W.C. (2005) Radiation damage effects in candidate titanates for Pu disposition: Pyrochlore. *J. Nucl. Mater.* **345**(2–3), 109–135.

Strachan D.M., Scheele R.D., Buck E.C., Kozelisky A.E., Sell R.L., Elovich R.J. and Buchmiller W.C. (2008) Radiation damage effects in candidate titanates for Pu disposition: Zirconoilite. *J. Nucl. Mater.* **372**, 16–31.

Wachtman J.B. (1996) *Mechanical properties of ceramics*, Wiley, New York.

Zamoryanskaya M.V., Burakov B.E., Bogdanov R.R. and Sergeev A.S. (2002) A cathodoluminescence investigation of pyrochlore, $(Ca,Gd,Hf,U,Pu)_2Ti_2O_7$, doped with ^{238}Pu and ^{239}Pu. in *Scientific Basis for Nuclear Waste Management XXV, Mater. Res. Society Symp. Proc.* **713**, 481–485.

Zamoryanskaya M.V., Konnikov S.G. and Zamoryanskii A.N. (2004) High-sensitivity system for cathodoluminescent studies with the Camebax electron probe microanalyzer. *Instrum. Exp. Tech.* **47**(4), 477–483.

Zamoryanskaya M.V. and Burakov B.E. (2004) Electron microprobe investigation of Ti-pyrochlore doped with Pu-238. in *Scientific Basis for Nuclear Waste Management XXVIII, Mater. Res. Society Symp. Proc.* **824**, 231–236.

Zamoryanskaya M.V. and Burakov B.E. (2006) Cathodoluminescence of actinide ions in crystalline host phases. in *Proceedings of the 8th Actinide Conference, ACTINIDES 2005, University of Manchester, UK*, 4–8 July 2005. Published by The Royal Society of Chemistry, pp. 767–769.

Chapter 6

Radiation Damage

Radiation damage and effects in materials have been studied extensively (Holmes-Siedle and Adams, 2002). Initially, metals were examined (Thompson, 1969) and later work focused more on inorganic
materials as summarised by Lehmann (1977). Lee (1983) summarised
the early theory of radiation damage including production of point
defects during irradiation, the nucleation and growth of defect clus-
ters and the evolution of the defect microstructure highlighting in
sapphire formation of dislocations from aggregation of interstitials
and bubbles and voids from aggregation of vacancies with, for example, He and Ar gas (Lee et $al.$, 1985). Radiation effects in metals were
discussed more recently by Was (2007). Experimental techniques to
examine radiation damage include:

(i) Characterising samples taken from reactor or other facility (e.g.,
irradiated in $situ$ by the full spectrum of radiation: neutron, α,
β and γ);

(ii) Characterising samples irradiated by ion accelerator ex $situ$;

(iii) Irradiating samples in a transmission electron microscope using
high-energy electrons or ions from a separate accelerator attach-
ed to the microscope (tandem facilities);

(iv) Incorporating radionuclides into the material during processing
whose radiation causes in $situ$ or self-irradiation.

158   *Crystalline Materials for Actinide Immobilisation*

Each of these techniques has benefits and drawbacks but all are experimentally challenging and expensive. As a result, modelling and simulation techniques have evolved to a significant level to support empirical data and reduce the number of active experiments needed.

Computer modelling has been used extensively by the nuclear community over the last 30 years. Common techniques include energy minimisation (Burnsall, 1979; Leslie, 1982; Grimes and Catlow, 1991), molecular dynamics (Parfitt and Grimes, 2008) and first principles simulations (Kotomin *et al.*, 2009). For example, the energy minimisation code "CASCADE" (Burnsall, 1979), which was developed at Harwell Laboratory in the UK, has been used extensively to simulate defect processes in ceramic materials (Busker *et al.*, 1999). The molecular dynamics method is an efficient technique that can predict the diffusion processes of ions, usually at elevated temperatures, and can also predict the mechanism of diffusion, which is difficult to access experimentally.

It is important to improve the performance of materials under extreme conditions such as under irradiation and at high temperature and pressure. Simulation techniques have been used in conjunction with experiment to study radiation tolerant ceramics (Sickafus *et al.*, 2000). An advantage of simulation studies is that they can lead to a deeper understanding of the origin of radiation damage and provide routes to design radiation tolerant ceramics (Sickafus *et al.*, 2007). For example, it has been predicted that radiation tolerance in pyrochlore materials can be correlated to the ease with which disordering effects are accommodated (Sickafus *et al.*, 2000). Ceramics in which defects can easily form are better at resisting radiation-induced amorphisation (Sickafus *et al.*, 2000). Such studies can highlight potential hosts for actinides and radioactive waste materials. With atomistic simulation techniques a wide range of compositions can be investigated thus leading the experimental studies to the more promising systems.

Recently, the behaviour of nanocomposites under irradiation has been investigated (Demkowicz *et al.*, 2008). Interfaces were predicted to act as sinks for radiation-induced defects and implanted species like helium (Demkowicz *et al.*, 2008; Misra *et al.*, 2007). Although most of the modelling work has been performed on metal-metal composites (Demkowicz *et al.*, 2008; Misra *et al.*, 2007), it is important to note that maximising the interface area per unit volume can result in a radiation-resistant material.

In a reactor, insoluble gas atoms such as krypton and xenon are typically formed in uranium dioxide due to the fission process. Additionally, helium gas is an issue for mixed oxide (MOX) fuels, inert matrix fuels (IMF) with minor actinides or during the long-term storage of fuel material. These gas atoms form microscopic bubbles within uranium dioxide (Nixon and Macines, 1981). The accumulation of the krypton, xenon and helium gas bubbles at grain boundaries leads to a degradation of the fuel performance (Ferry *et al.*, 2006). Through the use of classical molecular dynamics the interaction of radiation damage due to alpha decay with fission gas bubbles, which is a major resolution mechanism for both inter-granular and intra-granular gas bubbles, has been investigated (Parfitt and Grimes, 2008; Parfitt and Grimes, 2009). Importantly the mechanisms underpinning these interactions can have significant implications for the modelling of fuel performance at high burn-up and for MOX fuels (Parfitt and Grimes, 2008; Parfitt and Grimes, 2009).

Decay of radionuclides incorporated into the material causes *in situ* or self-irradiation. The long-term behaviour of crystalline actinide host-phases under alpha self-irradiation is difficult to simulate. Radiation damage effects have been examined in so-called natural analogue minerals showing metamict behaviour, i.e., becoming amorphous due to irradiation (Ewing *et al.*, 1987). These minerals naturally contain radionuclides and accumulate high doses of self-irradiation, causing the transition from the crystalline to disordered state. Radiation damage is often characterised by XRD and electron diffraction and TEM methods (see Figs. 6.1.2, 6.2.4, 6.2.5, 6.2.6 and 6.2.9). In addition, degradation of crystalline structure may be observed by cathodoluminescence (Fig. 5.3.2) and nuclear magnetic resonance (NMR) (Farnan *et al.*, 2007). Damage of crystalline structure is usually, but not always accompanied by reduced chemical

160   *Crystalline Materials for Actinide Immobilisation*

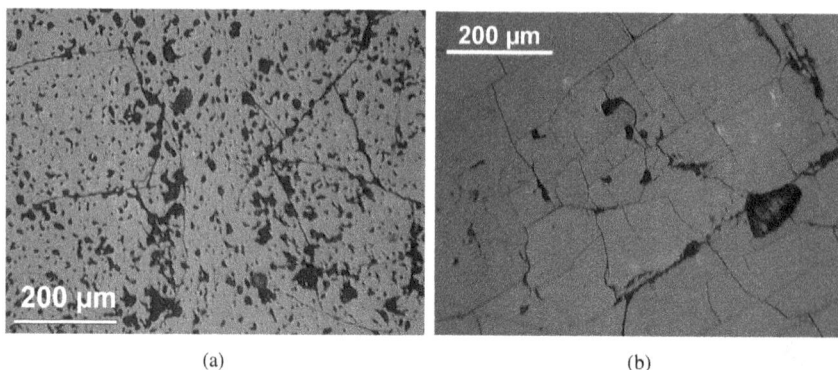

(a)   (b)

Fig. 6.1.1.   (a) Reflected light images of ^{238}Pu-doped pyrochlore-based ceramic, $(Ca,Gd,Pu,U,Hf)_2 Ti_2 O_7$ (8.7 wt.% ^{238}Pu), after cumulative dose of $26 \cdot 10^{23}$ alpha-decays/m^3; and (b) zircon single crystal (2.4 wt.% ^{238}Pu as average content) after cumulative dose of $7.5 \cdot 10^{17}$ alpha-decay/g.

durability (Table 5.6.2), swelling (density decrease) and crack formation in poly- and single-phase crystalline ceramics (Figs. 5.2.5, 5.4.1, 5.4.2 and 6.1.1). In some cases crack formation in crystalline materials is similar to that found in glassy wasteforms at high doses of alpha self-irradiation (Weber *et al.*, 1979).

Computer simulation of structural changes in zircon after just single alpha-decay events revealed the complexity of radiation damage processes leading to polymerisation, shear deformation and unit-cell expansion (Trachenko *et al.*, 2002).

A detailed review of radiation effects in crystalline ceramics has been published (Weber *et al.*, 1998). Alpha-decay releases alpha-particles with energies typically in the range of 4.5–5.8 MeV, and recoil nuclei with energies of 70 to 100 keV (Ewing *et al.*, 2004). Alpha-particles and alpha-recoils interact with the atomic structure causing radiation damage. An alpha-particle predominantly deposits its energy by ionisation processes, while the alpha-recoil loses most of its energy in elastic collisions with the nuclei of atoms in the solid (Ewing *et al.*, 2004). Cumulative irradiation dose can be quantified in *alpha-decays/m*^{3} in order to compare polycrystalline materials or in *alpha-decays/g* — for dense single-phase compounds, including single crystals. Another common unit for cumulative dose is *dpa* (displacements per atom). The calculated dose in displacements per atom is derived from:

$$Dose \cdot (dpa) = \frac{1,500 \cdot D \cdot M}{N_f N_A},\qquad (6.1)$$

where 1,500 is the average collisions of alpha-recoils with atoms after a single alpha-decay, D — total number of alpha-decays; M — molecular mass of the compound; N_f — total number of atoms in chemical formula; N_A — Avogadro's number.

Under irradiation at room temperature (depending on cumulative dose and energy) the original crystalline structure may:

1. be retained (monoclinic, tetragonal and cubic zirconia, $(Zr,...)O_2$; some monazite species, $(Ce,La,Eu,...)PO_4$);
2. be converted into another type of crystalline structure (e.g., pyrochlore $Gd_2Zr_2O_7$ can transform to a fluorite-type cubic structure (Wang *et al.*, 1999); or
3. become "amorphous" or "metamict" (as has been observed in zircon; Ti-pyrochlore; zirconolite; Pu-monazite, $PuPO_4$; xenotime, YPO_4; aluminate and ferrate garnet; murataite).

Increased temperature (during irradiation) suppresses the effect of radiation damage and supports recovery of the damaged structure, as does post-irradiation heat treatment (annealing). Complete degradation of the crystalline structure will not occur if the amorphisation rate is less or equal to the damage recovery rate (Ewing *et al.*, 2004). For each crystalline compound there is a critical temperature T_c above which amorphisation as a result of radiation damage does not occur. For some monazite and apatite species T_c is close to ambient temperature, leading to rapid thermal annealing of defects and recovery of the structure from any radiation damage.

Damage recovery may also be caused by irradiation-assisted processes as demonstrated by recrystallisation of amorphous fluorapatite (Meldrum *et al.*, 1997a); La-monazite; $ScPO_4$ and zircon (Meldrum *et al.*, 1997b) under an electron beam. The unusual

long-term stability and chemical durability of natural solid U-bearing
Zr-silicate gel (see Sec. 1.4.14) is assumed to be caused by two competing processes:

1. crystallisation of the gel into U-doped zircon assisted by self-
   irradiation and
2. metamictisation (amorphisation) of the crystallised zircon back to
   a gel-like state (Burakov *et al.*, 2006).

Radiation damage processes differ significantly for highly radioactive actinide-doped materials (solid solutions containing from several
to many wt.% Pu, Am, Np, Cm) compared with:

— undoped phases (externally irradiated by heavy ions simulating
   alpha-irradiation);
— host-phases doped with small amounts of highly radioactive
   actinides (less 0.5 wt.%);
— host-phases doped with comparable amounts of low radioactive
   actinides (U, Th).

Thus, it is very difficult to empirically verify the radiation effects
in actual wasteforms using such techniques. This difficulty is demonstrated by a study of natural and artificial zircon. The degradation
of the crystal structure of artificial zircon, $(Zr,Pu)SiO_4$, doped with
several wt.% ^{238}Pu takes place after 6–10 years at ambient conditions
and is similar to amorphisation of natural zircon, $ZrSiO_4$(containing
10^{-2}–10^{-1} wt.% U and Th), which achieves a similar dose after millions of years. A small difference is observed only in the early stages of
damage accumulation potentially arising from annealing of defects in
natural zircon (Murakami *et al.*, 1991). However, recovery of the crystal structure of highly-damaged amorphous samples of ^{238}Pu-doped
zircon compared with recovery of amorphous natural zircon at comparable cumulative dose is completely different under hydrothermal
conditions (Geisler*et al.*, 2005).

Another uncertainty is related to the critical actinide content
in the host-phase which affects the resistance of the solid solution
crystal structure to self-irradiation (Polezhaev, 1974). For instance,

samples of natural monazite, $(Ce,La,Th,U)PO_4$, containing at least 10 wt.% of Th and U and which have accumulated high self-irradiation doses are usually crystalline. This is consistent with the high radiation resistance of artificial monazite, $(La,Pu)PO_4$, doped with 8 wt.% ^{238}Pu after comparable radiation doses (Burakov *et al.*, 2004a). At the same time Pu-monazite, $PuPO_4$, is unstable under self-irradiation and becomes amorphous at a relatively low cumulative dose (Burakov *et al.*, 2004a).

Studies of zircon and monazite reveal that radiation damage effects depend not only on the intrinsic features of the crystal structures and cumulative doses but also on the chemical composition of the solid solutions. Ion-irradiations of undoped pyrochlore-structured $Gd_2Ti_2O_7$ and $Gd_2Zr_2O_7$ show different behaviour: finally, the titanate becomes amorphous but the zirconate remains crystalline although it changes structure to fluorite-type (Wang *et al.*, 1999). However, equivalent behaviour of actinide-doped Ti- and Zr-pyrochlore under alpha-decay is not guaranteed when compared to their undoped analogues under ion-irradiation. Alpha self-irradiation of Ti-pyrochlore doped with actinides destroys the solid solution causing formation of new actinide-bearing phases before final amorphisation of the pyrochlore phase (Zamoryanskaya and Burakov, 2004). Ingrowth of zirconolite has been observed in ^{238}Pu-doped Ti-pyrochlore ceramics stored at 250°C before both phases eventually become amorphous (Strachan *et al.*, 2005). Such processes are not well understood. Conversion of actinide-doped Zr-pyrochlore into the cubic fluorite-structure might also be accompanied by formation of separate actinide phases. There is some evidence from study of natural partially-amorphous titanite, $CaTiSiO_5$ (containing 0.04–0.08 wt.% U and Th) that self-irradiation caused the co-existence of two titanite phases with the same structure but different defect and radioactive impurity contents (Chrosch *et al.*, 1998).

Radiation damage may lead to a change of actinide valence state (Zamoryanskaya *et al.*, 2002) that also affects the solid solution crystal structure and phase stability. The process of helium accumulation in the matrix of actinide-doped materials that have received high doses of self-irradiation may affect the materials' mechanical durability. In addition, decay of actinides is accompanied by formation of daughter elements which may have completely different effects on the host than the initial actinide.

In this chapter we now review two common methods of simulating radiation damage: external ion-irradiation and internal doping with short-lived isotopes such as ^{238}Pu (half-life = 87.7 years) and ^{244}Cm (half-life = 18.1 years).

6.1. Ion-irradiation

Ion-irradiation of inactive or low-activity crystalline materials by ions such as Ar^+, Kr^{2+}, Xe^+; He^+ and Au^{3+} is well established and allows evaluation of the materials' resistance to (simulated) alpha-radiation damage. In general, this method is not used to investigate macro-scale samples of ceramics and single crystals but just applied to small (mm) samples. Irradiated samples are typically examined by high-resolution transmission electron microscopy (HRTEM) in combination with electron diffraction (Fig. 6.1.2). Conversion from fully crystalline (Fig. 6.1.2(a)) to a completely amorphous state (Fig. 6.1.2(d))) through some intermediate stages (Figs. 6.1.2(b) and 6.1.2(c)) is easily observed. Ion-irradiation might also be carried out at different temperatures enabling the critical temperature of amorphisation (T_c) to be identified.

Ion-irradiations have been performed on many ceramics including zircon and hafnon (Weber *et al.*, 1994); $ThSiO_4$ (Meldrum *et al.*, 1999); phosphates of La, Pr, Nd, Sm, Eu and Gd with monazite structure and Sc, Y, Tb, Tm, Yb and Lu phosphates with zircon structure (Meldrum *et al.*, 1997c); pyrochlore solid solutions, $Gd_2(Ti,Zr)_2O_7$, including $Gd_2Ti_2O_7$ and $Gd_2Zr_2O_7$ end-members (Wang *et al.*, 1999); cubic and momoclinic zirconia (Degueldre *et al.*, 1997; Sickafus *et al.*, 1999; Wang *et al.*, 2000); murataite (Lian *et al.*, 2005); silicate apatite, $Ca_2La_8(SiO_4)_6O_2$ (Wang and Weber, 1999); phospho-silicate apatite, $Ca_7Nd_3(SiO_4)_3(PO_4)_3F_2$,

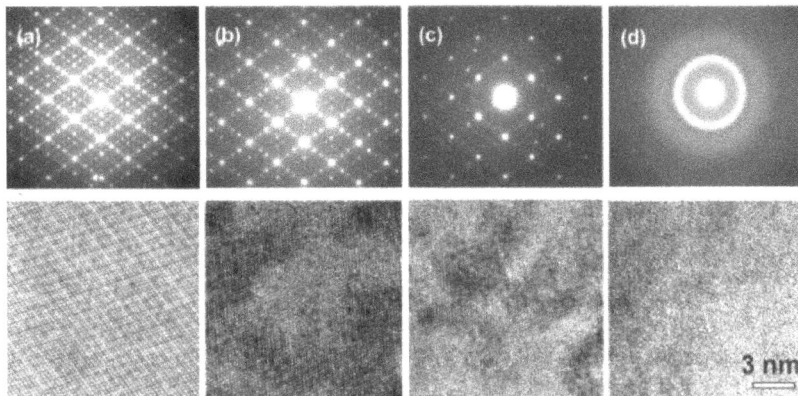

Fig. 6.1.2. Electron diffraction (top row) and high-resolution TEM images (lower row) of five-multiple murataite, $(Ca,Mn,U)_5(Ti,Zr)_8(Al,Fe)_2O_{27}$ samples, after ion-beam irradiation by Kr^{2+} at $25°C$ and different doses in ions/cm^2 x 10^{13} (and in dpa): (a) initial unirradiated sample; (b) 4.38 (0.035); (c) 9.39 (0.075); (d) 15.6 (0.125) (Lian *et al.*, 2005). Courtesy of Dr. S.V. Yudintsev of the Institute of Geology of Ore Deposits (IGEM), Russia.

and $Ca_9Nd(SiO_4)(PO_4)_5F_2$ (Soulet *et al.*, 2001); fluorapatite, $Ca_{10}(PO_4)_6F_2$, and hydroxyapatite, $Ca_{10}(PO_4)_6(OH)_2$ (Soulet *et al.*, 2001); TPD, $Th_4(PO_4)_4P_2O_7$ (Pichot *et al.*, 2001); silicate and Zr-ferrate garnet species (Utsunomiya *et al.*, 2002).

The main findings of ion-irradiation studies are:

— Radiation damage processes in crystalline materials usually include several stages, even in chemically simple compounds;
— Cubic and monoclinic zirconia as well as zirconates with pyrochlore structure have high resistance to radiation damage;
— Although different prospective (undoped) actinide host-phases demonstrate variable resistance under ion-irradiation, these data cannot be used directly to distinguish the most reliable host-phase for actinide immobilisation;
— No information from this method is available to understand the stability of actual highly-radioactive actinide solid solutions under radiation damage.

6.2. Doping with ^{238}Pu and ^{244}Cm

The radioactivity of ^{238}Pu is ~270 times higher than ^{239}Pu; and 24,000 times higher than ^{237}Np. Curium-244 is 24 times more radioactive than ^{241}Am and 400 times more than ^{243}Am (see Table 1.1.2). The use of ^{238}Pu and ^{244}Cm allows accelerated radiation damage experiments, which simulate the long-term behaviour of host-phases doped with tetra- and tri-valent long-lived actinides. The main advantage of this method is that macro-samples (ceramic pellets and single crystals from 1 mm to several cm in size) of actual actinide solid solutions at high cumulative doses of self-irradiation are examined. A disadvantage of this method is the practical problem of handling significant amounts of ^{238}Pu and ^{244}Cm. Just a few laboratories in the world have licensed facilities to work with such isotopes at macro-scale, e.g., from tens to hundreds milligrams.

The next few sections summarise results from such accelerated radiation damage studies of ceramics.

6.2.1. *Zircon/zirconia and hafnon/hafnia ceramics*

Polycrystalline zircon, $(Zr,Pu)SiO_4$, (density ~4.7 g/cm^3) doped with 10 wt.% Pu (all isotopes) or 8.85 wt.% ^{238}Pu was synthesised and studied by XRD at PNNL (Exharos, 1984; Weber, 1991). This sample was later studied by neutron diffraction (Fortner *et al.*, 1999) and NMR (Farnan *et al.*, 2007). Results of SEM and EPMA were not reported. This zircon became fully X-ray amorphous after a dose of $6.7 \cdot 10^{18}$ alpha-decays/g or roughly $300 \cdot 10^{23}$ alpha-decays/m^3. Neutron diffraction suggested the local structural units of zircon are preserved in the fully amorphous material, with little phase separation or recrystallisation.

Zircon/zirconia, $(Zr,Pu)SiO_4/(Zr,Pu)O_2$, ceramic (density 4.4 g/cm^3) doped with ^{238}Pu was fabricated at KRI (Burakov *et al.*, 2001; Geisler *et al.*, 2005). XRD revealed this sample consisted of approximately 85 wt.% zircon and 15 wt.% tetragonal zirconia. The bulk Pu content (all isotopes) in the zircon measured by EPMA was 5.7 wt.% corresponding to 4.6–4.7 wt.% ^{238}Pu. The Pu content in the

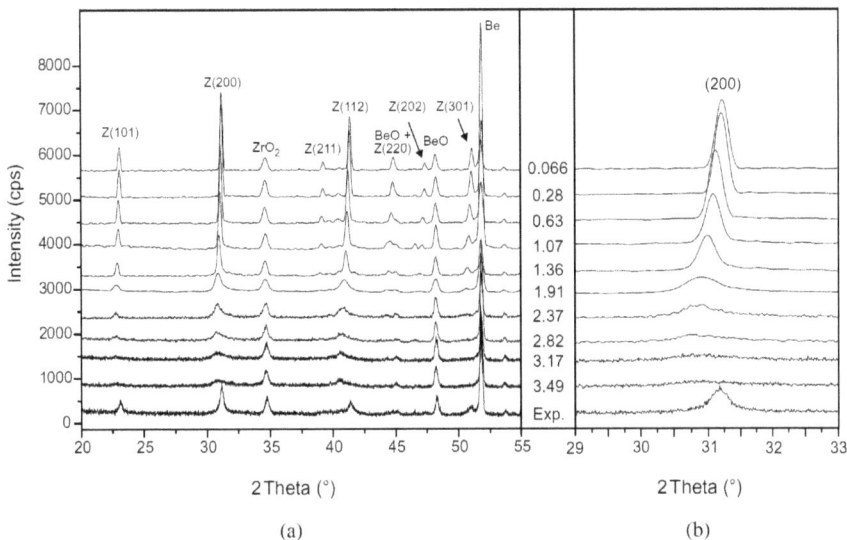

Fig. 6.2.1.   XRD of ^{238}Pu-doped zircon/zirconia ceramic after accumulation of different self-irradiation doses (given in 10^{18} alpha-decays/g) and after hydrothermal treatment at 175°C (labelled "Exp".): (a) whole spectra and (b) details of the (200) peak of zircon. Diffraction peaks from zircon (Z), tetragonal zirconia (ZrO_2) and from the Be window (Be and BeO) are indicated (Geisler *et al.*, 2005).

tetragonal zirconia was not measured because grain sizes were too small (1–3 μm) for precise EPMA. The zircon became X-ray amorphous after a dose of roughly $3.9 \cdot 10^{18}$ alpha-decays/g or $190 \cdot 10^{23}$ alpha-decays/m^3, although the tetragonal zirconia preserved its initial crystallinity (Fig. 6.2.1). After zircon amorphisation the ceramic pellet remained mechanically durable and free of cracks. Geometrical measurements did not reveal any density change. It is possible any swelling effect was accommodated by the ceramic porosity. This pellet was then used to examine damage recovery under hydrothermal conditions at 175°C (Geisler *et al.*, 2005). Partial recovery of the zircon crystal structure (Fig. 6.2.1 — spectrum labelled Exp.) did not affect the ceramic mechanical durability. A separate ceramic pellet was used for repeated static leach tests and demonstrated chemical durability after different cumulative doses (Table 5.2.1).

Samples of ^{238}Pu-doped zircon and hafnon ceramics fabricated at KRI in 2001 were studied by repeated XRD analysis. The samples

168   *Crystalline Materials for Actinide Immobilisation*

Table 6.2.1.   Expansion of unit-cell parameters of [238]Pu-doped zircon and hafnon (from direct XRD analysis of ceramic pellets without sample grinding) as a function of cumulative dose. Approximate content of [238]Pu in each sample is 4.7 wt.%.

Zircon		Hafnon	
Cumulative dose, $(10^{23}$ alpha-decays/ $m^3)$	Unit-cell parameters (error), (Å)	Cumulative dose, $(10^{23}$ alpha-decays/$m^3)$	Unit-cell parameters (error), (Å)
1.4	a = 6.634(1) c = 6.001(1)	2	a = 6.608(1) c = 5.996 (1)
14	a = 6.657(2) c = 6.029(4)	19	a = 6.629(1) c = 6.020(1)
22	a = 6.664(2) c = 6.039(4)	30	a = 6.643(1) c = 6.035(1)
39	a = 6.678(2) c = 6.045(4)	53	a = 6.671(2) c = 6.034(4)
53	a = 6.716(2) c = 6.084(4)	72	a = 6.677(?) c = 6.075(?)
66	a = 6.720(?) c = 6.091(?)	91	a = 6.646(?) c = 6.074(?)
77	a = 6.951(?) c = 6.039(?)	105	Impossible to measure

contained admixtures of 3–5 wt.% minor tetragonal zirconia, $(Zr,Pu)O_2$, or hafnia, $(Hf,Pu)O_2$, respectively. Thin (0.5–1.0 mm) ceramic pellets used for XRD analysis (without grinding) had densities of 3.7 g/cm^3 for zircon and 5.1 g/cm^3 for hafnon. Similar changes of unit-cell parameters of zircon and hafnon as a function of cumulative dose are shown in Table 6.2.1. Six years after the ceramic synthesis, when both phases have become X-ray amorphous, a pellet of zircon ceramic was extracted from hermetic cassette with Be-window and examined under an optical microscope. No cracks were observed (Fig. 6.2.2).

6.2.2. *Zircon single crystal*

Single crystals of zircon doped with [238]Pu (Figs. 5.2.3, 5.2.5, 5.4.1, 5.4.2 and 6.1.1(b)) were grown by the flux method at KRI in 2001

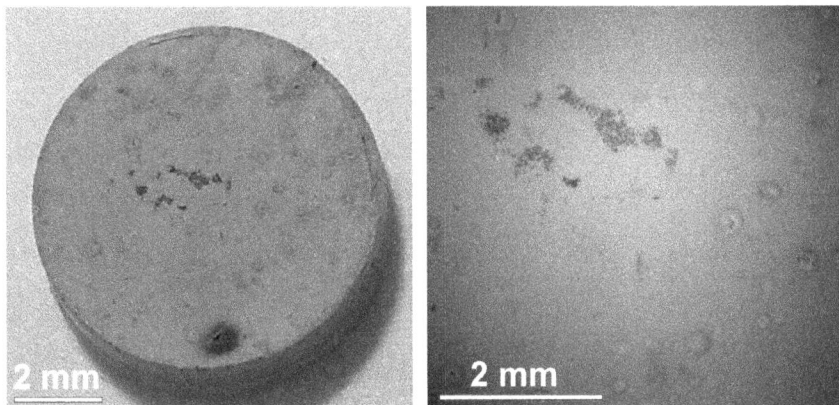

Fig. 6.2.2.   Pellet of ^{238}Pu-doped (approximately 5 wt.% ^{238}Pu) zircon ceramic six years after fabrication (cumulative dose approximately $200 \cdot 10^{23}$ alpha-decays/m^3). Zircon (according to XRD) has become amorphous. However, the ceramic matrix remains intact and free of cracks.

(Hanchar *et al.*, 2003; Burakov *et al.*, 2008). The average bulk content of ^{238}Pu measured by gamma-spectrometry in selected separate crystals varied from 2.2 to 2.6 wt.%. 2.4 wt.% ^{238}Pu was thus used as the average content for all calculations of cumulative doses. The distribution of Pu measured by EPMA (for all Pu isotopes) was not homogeneous (Fig. 5.2.4 and Table 5.2.1). Crystals glowed in the dark in the first months after synthesis (Fig. 2.4.1(a)). One large crystal (Fig. 5.4.1) was placed in a special glass-cassette (Fig. 5.4.3), allowing repeated optical microscope observation over many years. Self-irradiation caused crystal cracking and a change of colour from initial pink-brown to brown, then to grey-brown, yellow-grey and green-grey. Formation of tiny unidentified particles near the crystal was observed (Fig. 5.4.1), possibly arising from fracture of the crystal surface under self-irradiation. This process could be due to inhomogeneous Pu distribution, particularly near the outer edges of the crystal (Table 5.2.1). Inhomogeneous distribution of actinides in non-conducting host materials may lead to mechanical destruction due to accumulation of electric charge (Kachalov *et al.*, 1987; Ojovan and Poluektov, 2001). At the present time (2009) this crystal is stored in a cassette and no further cracking has been observed.

Table 6.2.2. Zircon unit-cell parameters as a function of cumulative dose. XRD of ground single crystal doped with (on average) 2.4 wt.% ^{238}Pu.

Cumulative dose, (10^{17} alpha-decays/g)	Unit-cell parameters, (Å)	
	a	c
0 (immediately after synthesis)	No data	No data
6.3	6.616(2)	6.006(3)
8.8	6.620(2)	6.013(3)
10	6.617(2)	5.987(3)
12	6.620(2)	6.008(3)
13	6.620(2)	6.017(3)
15	6.623(2)	6.011(3)
18	6.615(2)	6.012(3)
Synthetic undoped ZrSiO$_4$, JCPDS standard 6-626	6.604	5.979

Selected crystals were ground with crystalline α-SiO$_2$ powder (as internal standard) and the resulting homogeneous powder sealed in a hermetic cassette with Be-window (Fig. 5.1.1), which enabled precise repeated XRD analysis. The variation of zircon unit-cell parameters as a function of cumulative dose is shown in Table 6.2.2.

6.2.3. *Cubic zirconia ceramic*

Samples of single-phase gadolinia-stabilised cubic zirconia doped with ^{238}Pu (Fig. 6.2.3) have been synthesised at KRI (Burakov *et al.*, 2002a; Burakov *et al.*, 2002b; Burakov *et al.*, 2004b). EPMA indicated this ceramic was $Zr_{0.79}Gd_{0.14}Pu_{0.07}O_{1.99}$. The ^{238}Pu content was estimated to be ~9.9 wt.% and the ceramic density was 5.6 g/cm^3. Neither matrix swelling nor cracking were observed after a cumulative dose of $509{\cdot}10^{23}$ alpha-decays/m^3 (or $8.7{\cdot}10^{18}$ alpha-decays/g). At this dose cubic zirconia retained its fluorite-type crystal structure and no new inclusions of Pu phases were found by SEM in the ceramic.

Detailed XRD analysis revealed unusual behaviour of the main (111) reflection of cubic zirconia depending on cumulative dose.

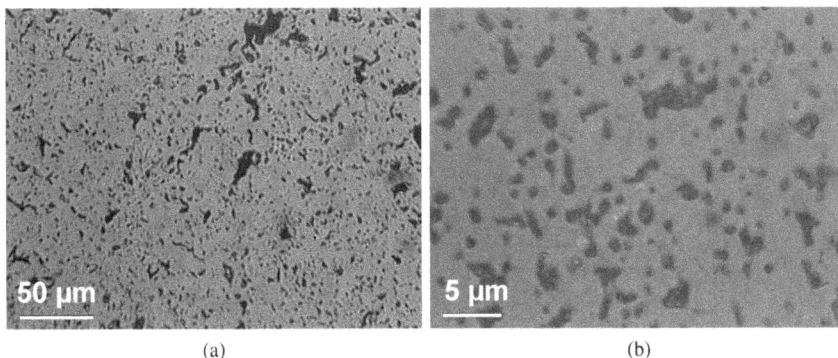

(a) (b)

Fig. 6.2.3.   Reflected light image of single-phase cubic zirconia ceramic, $Zr_{0.79}Gd_{0.14}Pu_{0.07}O_{1.99}$, doped with approximately 9.9 wt.% ^{238}Pu (Burakov *et al.*, 2002). No change in matrix microstructure or crack formation was observed at cumulative dose $509 \cdot 10^{23}$ alpha-decays/m^3.

After doses of (in alpha-decay/m$^3 \cdot 10^{23}$): 3, 27, 62 and 110, this peak looked similar and fairly broad. However, after doses of 134, 188, 234 and 277, it became narrower with increased intensity (Fig. 6.2.4). Self-irradiation of cubic zirconia by alpha-particles appears to be accompanied by two processes: accumulation of defects in the crystal structure and repeated self-annealing of these defects under ambient conditions.

Normalised Pu mass loss (NL_{Pu}) from the zirconia matrix (in deionised water at 90°C, during a 28-day leach test) increased from 0.04 g/m^2 after $11 \cdot 10^{23}$ alpha-decay/m^3 to 0.35 and 0.37 g/m^2 after 56 and $81 \cdot 10^{23}$ alpha-decay/m^3, respectively (Table 5.6.2). However, further self-irradiation of cubic zirconia to dose levels of $127 \cdot 10^{23}$ alpha-decay/m^3 was characterised by decrease of NL_{Pu} to 0.24 g/m^2.

6.2.4. *Monazite ceramic*

^{238}Pu-doped monazite ceramics were fabricated at KRI (Burakov *et al.*, 2004a). One sample (with density 4.7 g/cm^3) was single phase, (La,Pu)PO$_4$, containing 8.1 wt.% ^{238}Pu. Another sample (with density 4.9 g/cm^3) consisted of ~70–80 wt.% PuPO$_4$ with monazite structure, and a minor phase, PuP$_2$O$_7$. The bulk ^{238}Pu content in this ceramic measured by gamma-spectrometry was 7.2 wt.%.

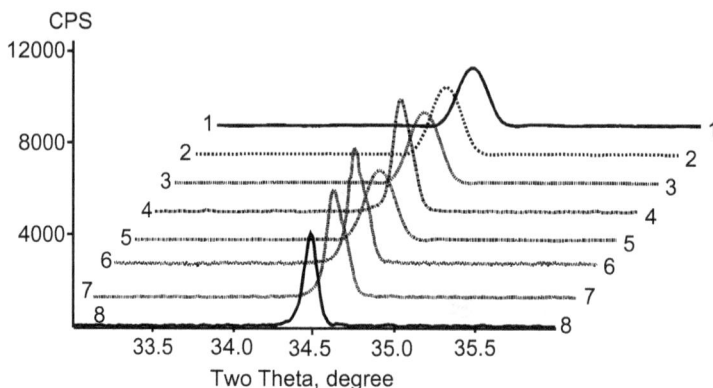

Fig. 6.2.4.   Behaviour of (111) XRD reflection of gadolinia-stabilised cubic zirconia doped with 9.9 wt.% ^{238}Pu after cumulative dose ($\cdot 10^{23}$ alpha-decay/m^3): (1) 3; (2) 27; (3) 62; (4) 110; (5) 134; (6) 188; (7) 234 and (8) 277 (Burakov *et al.*, 2004b). CPS is counts per second.

The (La,Pu)PO$_4$ monazite remained crystalline to a cumulative dose of $119 \cdot 10^{23}$ alpha decays/m^3 (Fig. 6.2.5). However, the self-irradiation did cause some changes in intensity and width of the monazite peaks, although the unit cell parameters did not vary as a function of cumulative dose.

In contrast to the La-monazite sample, the Pu-phosphate end-member sample became nearly amorphous at a relatively low cumulative dose of $42 \cdot 10^{23}$ alpha-decays/m^3 (Fig. 6.2.6). The minor phase in the Pu-phosphate ceramic, PuP$_2$O$_7$, became amorphous before the PuPO$_4$. Swelling and macrocrack formation resulting from self-irradiation damage were observed in the Pu-phosphate ceramic. Also, under self-irradiation this sample changed colour from deep blue to black. The La-phosphate was characterised by a similar change of colour from light blue to grey, although no swelling or crack formation have been observed to date.

6.2.5. *Monazite single crystal*

Single-crystal Eu-monazite, (Eu,Pu)PO$_4$, doped with ^{238}Pu was synthesised at KRI at the beginning of 2004 (Burakov *et al.*, 2008). The ^{238}Pu content measured by gamma-spectrometry was ∼4.9 wt.% (or

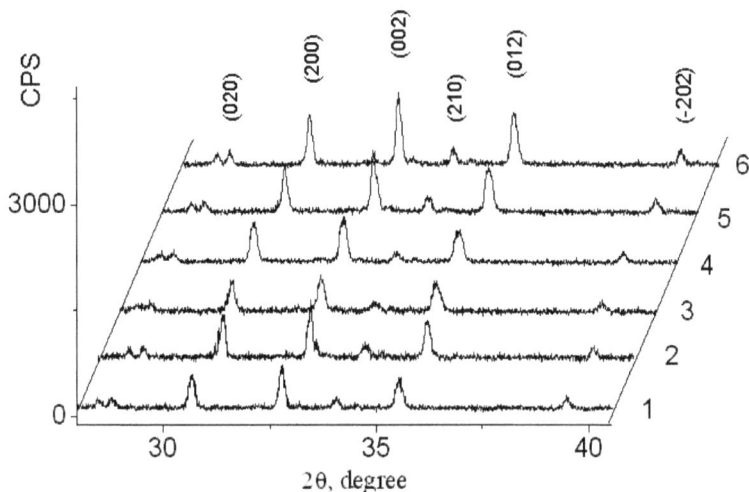

Fig. 6.2.5. XRD of La-monazite, $(La,Pu)PO_4$, doped with 8.1 wt.% ^{238}Pu as a function of cumulative dose ($\cdot 10^{23}$ alpha-decays/m^3): (1) 1.5; (2) 19; (3) 47; (4) 72; (5) 93; (6) 119 (Burakov *et al.*, 2004a).

~6 wt.% for all Pu isotopes). Selected crystals were placed in a sealed glass-cassette for repeated optical microscopy observation. Formation of dispersed non-identified particles around the crystals (Fig. 6.2.7) was clearly observed 14 months after crystal synthesis (after cumulative dose ~$1.1\cdot 10^{18}$ alpha-decays/g. Again, this likely arises from mechanical destruction of the crystal surface under self-irradiation. After $5.2\cdot 10^{18}$ alpha-decays/g (~64.5 months after crystal synthesis), the process of crystal surface destruction has become significant (Fig. 6.2.8). Initially the crystals were transparent with a homogeneous deep pink-purple colour. It is presently difficult to evaluate the change of colour as a result of self-irradiation because the crystal surface has become dull but it appears to be grey-purple.

6.2.6. *Ti-pyrochlore ceramic*

Polycrystalline pyrochlore, $(Gd,Cm)_2Ti_2O_7$, doped with ~3 wt.% ^{244}Cm was obtained in the 1980s at PNNL and studied by XRD, electron diffraction and TEM (Weber *et al.*, 1986), which revealed it became amorphous at a cumulative dose of $(170–190)\cdot 10^{23}$

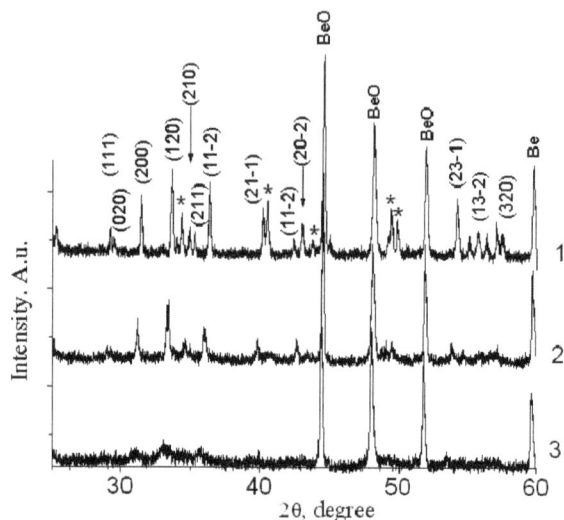

Fig. 6.2.6.   XRD of ceramic based on Pu-monazite, $PuPO_4$, containing 7.2 wt.% ^{238}Pu as a function of the cumulative dose ($\times 10^{23}$ alpha-decays/m^3): (1) 1.3; (2) 17 and (3) 42. Reflections from the minor phase, PuP_2O_7, are marked as (*). BeO and Be are reflections from the Be-window that covered the ceramic sample (Burakov *et al.*, 2004a).

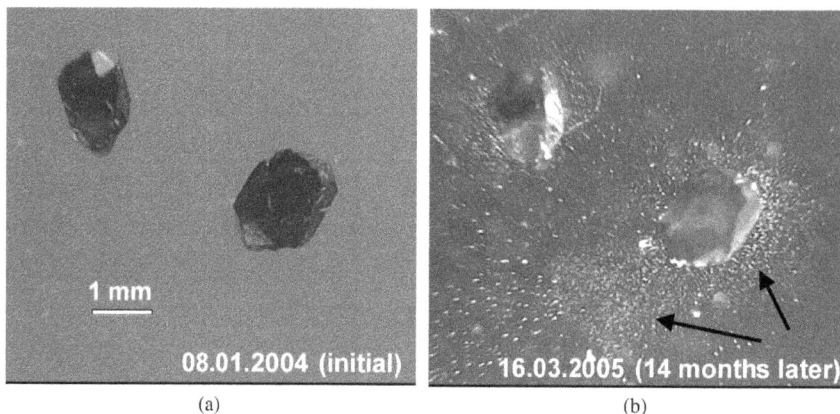

Fig. 6.2.7.   Single crystals of Eu-monazite, $(Eu,Pu)PO_4$, doped with 4.9 wt.% ^{238}Pu: (1) immediately after synthesis and (2) 14 months later (after $\sim 1.1 \cdot 10^{18}$ alpha-decays/g). Formation of dispersed particles around crystals is marked by arrows (Burakov *et al.*, 2008).

alpha-decay/m^3. The observed macroscopic swelling was 5.1%. Ceramic fracture toughness increased $\sim$1.5 times after a cumulative dose of $(70\text{--}80)\cdot10^{23}$ alpha-decay/m^3, however further self-irradiation caused it to decrease. Static leach tests (in deionised water, at 90°C for 14 days) revealed increased Cm and ^{240}Pu (decay product of ^{244}Cm) leaching as result of self-irradiation. Normalised mass loss (NL) increased (in g/m^2): from 0.01 to 0.17 for Cm and from 0.12 to 5.93 for Pu.

Intensive investigations of titanate pyrochlore-based ceramics doped with ^{238}Pu were carried out at PNNL (Strachan *et al.*, 2000; Strachan *et al.*, 2002; Strachan *et al.*, 2004; Strachan *et al.*, 2005). The starting precursor composition was (in wt.%): CaO — 12.06; Gd_2O_3 — 7.65; HfO_2 — 10.88; MoO_3 — 0.10; PuO_2 — 12.31; TiO_2 — 36.17; UO_2 — 20.82. The specimens were diphasic consisting of a pyrochlore with apparent chemical formula $Ca_{1.13}Gd_{0.22}Hf_{0.12}Pu_{0.24}$ $U_{0.40}(Ti_{1.90}Hf_{0.10})O_7$ and Hf-rutile, $(Hf_{0.1},Ti_{0.9})O_2$. The pyrochlore phase became X-ray amorphous at a cumulative dose of $(1.0\text{--}2.3)\cdot10^{18}$ alpha-decays/g. The geometrical volume swelling observed was 7.4% although the specimens remained intact. No evidence for microcracking was observed by SEM. It was reported that "radiation induced damage has no effect on the dissolution of this material" (Strachan *et al.*, 2004). Surprisingly, samples stored at 250°C revealed ingrowth of zirconolite (before the pyrochlore became amorphous) under self-irradiation.

Titanate pyrochlore ceramics doped with 8.7 wt.% ^{238}Pu have been fabricated at RIAR under a contract with LLNL (Volkov *et al.*, 2001). The target composition following LLNL was (in wt.%): CaO — 9.9; Gd_2O_3 — 7.8; HfO_2 — 10.6; TiO_2 — 35.4; UO_2 — 23.7; PuO_2 — 12.6. According to RIAR data the pyrcochlore became X-ray amorphous at a cumulative dose of $(120\text{--}140)\cdot10^{23}$ alpha-decays/m^3. Some RIAR samples were sent to KRI for a parallel study. Characterisation at the KRI only started 4 months after the ceramic synthesis so there is a lack of information on the initial ^{238}Pu-doped pyrochlore from XRD and EPMA. At a cumulative dose of $57\cdot10^{23}$ alpha-decays/m^3, the pyrochlore chemical composition was observed to be inhomogeneous by EPMA (in wt.% element,

176   *Crystalline Materials for Actinide Immobilisation*

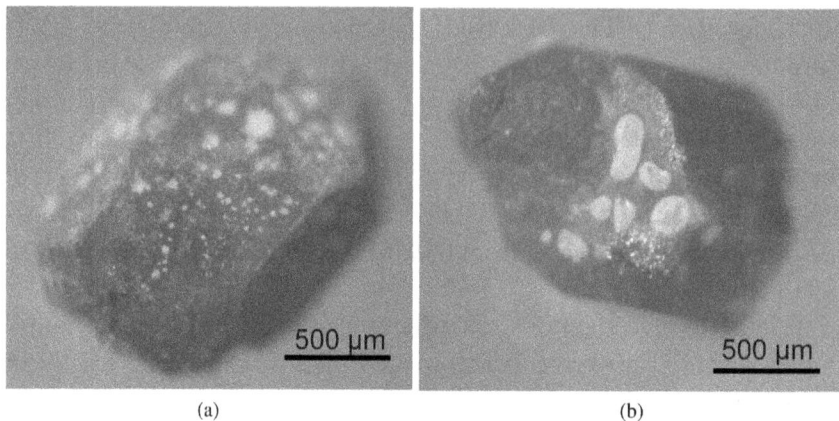

Fig. 6.2.8.   Surface destruction of single crystals of Eu-monazite, $(Eu,Pu)PO_4$, doped with 4.9 wt.% ^{238}Pu 64.5 months after synthesis at cumulative dose $\sim 5.2 \cdot 10^{18}$ alpha-decays/g.

"from-to"): Ca — 6.7–7.9; Hf — 3.6–7.1; Pu — 9.5–26.8; U — 13.2–32.9; Gd — 4.0–8.3; Ti — 19.9–21.5 (Burakov *et al.*, 2002a; Zamoryanskaya *et al.*, 2002). The estimated pyrochlore formula was $Ca_{0.9}(U_{0.3-0.5}Pu_{0.2-0.5}Gd_{0.1-0.2}Hf_{0.1-0.2})Ti_2O_7$. The pyrochlore became nearly completely amorphous after $(110–130) \cdot 10^{23}$ alpha-decays/m^3 (Fig. 6.2.9) accompanied by a decrease of ceramic density of ~ 10 % compared to the initial sample (from 4.8 to 4.3 g/cm^3). Inclusions of $(U,Pu)O_x$ and $(Hf,Ti,Ca)O_x$ were identified in the ceramic matrix by SEM and assumed to be newly formed (Zamoryanskaya *et al.*, 2002). Repeated SEM examination (about once every three months over several years) revealed the number of these inclusions increased with cumulative dose. This correlated with a shift of the main (222) and (400) pyrochlore peaks depending on cumulative dose to the low-angle direction (Fig. 6.2.9) corresponding to the position of the pure Ti-pyrochlore phase, $Gd_2Ti_2O_7$. It was suggested that self-irradiation might cause (before final pyrochlore amorphisation) partial destruction of the $(Ca,Gd,Hf,Pu,U)_2Ti_2O_7$ solid solution into different phases such as: Ti-pyrochlore with lower impurities, in particular, U and Pu; $(U,Pu)O_x$ and $(Hf,Ti,Ca)O_x$ (Zamoryanskaya *et al.*, 2002). Results of static leach tests at KRI (see Table 5.6.2) and RIAR (Lukinykh *et al.*, 2002) revealed a decrease in chemical

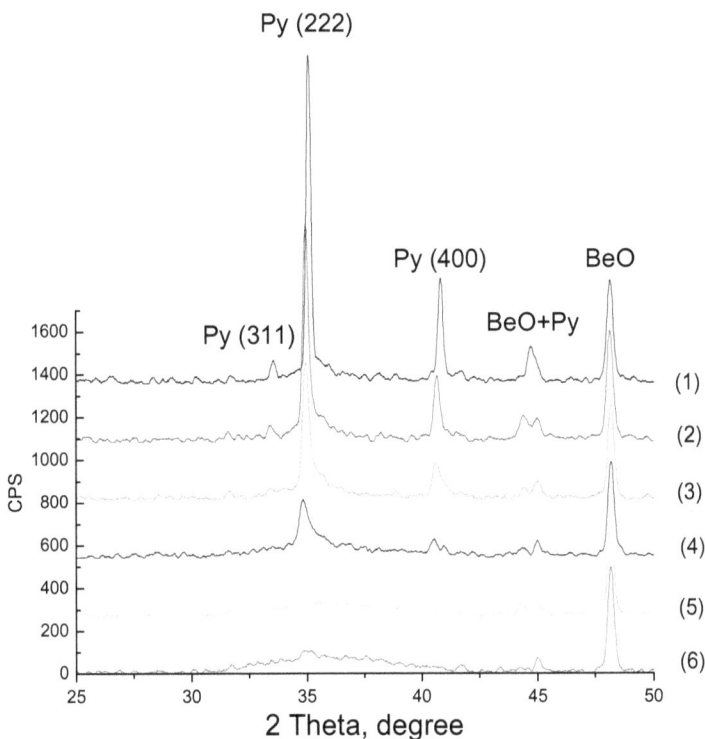

Fig. 6.2.9.   XRD from titanate pyrochlore-based ceramic doped with 8.7 wt.% ^{238}Pu after cumulative doses ($\times 10^{23}$ alpha-decays/m^3) of: (1) 26; (2) 43; (3) 57; (4) 82; (5) 110 and (6) 130. BeO — reflections from the Be window covering ceramic sample; BeO + Py — overlapping of BeO and pyrochlore peaks (Burakov and Anderson, 2002).

durability of the ceramic samples as a result of self-irradiation. These data contradict the PNNL conclusions (Strachan *et al.*, 2004). Cathodoluminescence indicated a change of uranium valence state in pyrochlore under irradiation damage and accumulation of uranyl-ion, $(UO_2)^{2+}$, and complex $(UO_4)^{2-}$ (Zamoryanskaya *et al.*, 2002).

6.2.7. *Zr-pyrochlore ceramic*

Californium zirconate pyrochlore, $Cf_2Zr_2O_7$, with unit-cell parameter $a = 10.63(2)$ Å was synthesised in France and studied by CEA and ORNL researchers using XRD (Raison *et al.*, 2002). ^{249}Cf with half-life 351 years was used. After six months of storage at a cumulative dose of $1.17 \cdot 10^{18}$ alpha-decays/g the pyrochlore reflections have almost disappeared and the crystalline structure converted to a cubic fluorite-type. It was assumed this process was accompanied by oxidation of Cf from (3+) to (4+) valence state.

Pyrochlore-based ceramic with a target composition $Gd_{1.935}$ $Cm_{0.065}TiZrO_7$ was synthesised at RIAR and studied in cooperation with IGEM and Moscow SIA "Radon" using XRD (Yudintsev *et al.*, 2009). The sample consisted of pyrochlore with unit-cell parameter $a = 10.328(1)$ Å and some minor phases: $ZrTiO_4$, monoclinic ZrO_2 and a perovskite-structured phase. Five hundred days after synthesis at a cumulative dose of $2.46 \cdot 10^{18}$ alpha-decays/g the pyrochlore unit-cell parameter increased to $10.399(10)$ Å. After a dose of $3.3 \cdot 10^{18}$ alpha-decays/g XRD patterns contained reflections mainly of two phases: the target Ti/Zr- pyrochlore and traces of zirconate pyrochlore, $Gd_2Zr_2O_7$, "whose reflections were observed even after complete amorphisation of the target phase". The target phase thus became X-ray amorphous $\sim$970 days after its preparation at a cumulative dose of $4.6 \cdot 10^{18}$ alpha-decays/g.

6.2.8. *Zirconolite ceramic*

XRD and TEM of a ^{244}Cm-doped ($\sim$3 wt.% ^{244}Cm) zirconolite, $Ca(Zr,Cm)Ti_2O_7$, ceramic made at PNNL (Weber *et al.*, 1986) revealed it became amorphous at a cumulative dose of $(210–230) \cdot 10^{23}$ alpha-decay/m^3. The volume swelling observed was 6%. As in the pyrochlore (see Sec. 6.2.6), the ceramics fracture toughness increased $\sim$1.5 times after a cumulative dose $(80–90) \cdot 10^{23}$ alpha-decay/m^3, although further self-irradiation caused it to decrease. Static leach testing (in deionised water, at 90°C for 14 days) revealed increased leaching of some elements as a result of self-irradiation. Normalised mass loss (NL) increased $\sim$8 times for Ca and 10 times for Pu (^{240}Pu is a decay product of ^{244}Cm). Surprisingly, the NL_{Cm} remained the same for crystalline and amorphous samples and was very low at 0.001 g/m^3.

Samples of ^{238}Pu and ^{244}Cm doped Synroc C were prepared in 1987 at the UKAEA Laboratories at Harwell (Hambley *et al.*, 2008). Examination after 20 years in 2007 (self-irradiation dose $\sim$1600·10^{23} alpha-decay/m^3) revealed microcracking only in actinide-free regions (such as crystalline hollandite, BaAl$_2$Ti$_8$O$_{16}$, and (Ti,Al)$_x$O$_{2-x}$ with structure similar to rutile, TiO$_2$). Precise quantitative measurement of actinide contents in amorphous zirconolite and perovskite has not been carried out, however, mention was made of varying Pu content from 1 to 3 at.% for zirconolite and from 2 to 4 at.% for perovskite.

Before the initial experiments at PNNL on synthesis of Cm-doped zirconolite samples of ^{238}Pu-substituted "cubic zirconolite (CaPuTi$_2$O$_7$)" with fluorite-type structure (although CaZrTi$_2$O$_7$ is monoclinic as well as Ca(Zr$_{0.8}$Pu$_{0.2}$)Ti$_2$O$_7$) were obtained at LANL (Clinard *et al.*, 1982). This material became X-ray amorphous by a dose of 130·10^{23} alpha-decay/m^3. Amorphisation caused swelling (volume increase of 2.2%) and formation of microcracks in the sample matrix (already observed after 24 days' storage) and in particular near the sample edge. Electron diffraction of the fully amorphous sample did not reveal any crystallites except PuO$_2$ (initial minor phase in ceramic sample) and TiO$_2$ (rutile). Synthesis of ^{238}Pu-doped monoclinic zirconolite, Ca(Zr$_{0.8}$Pu$_{0.2}$)Ti$_2$O$_7$, is mentioned by the same authors but without any detail.

A thorough study of ^{238}Pu-doped zirconolite-based ceramics was carried out at PNNL (Strachan *et al.*, 2002; Strachan *et al.*, 2004; Strachan *et al.*, 2008). Ceramic samples contained from 7.4 to 11.26 wt.% PuO$_2$. Zirconolite (the samples were mainly zirconolite phase) had complex chemical composition, which was calculated as Ca$_{0.86}$(Al$_{0.10}$Gd$_{0.05}$Hf$_{0.93}$Pu$_{0.13}$U$_{0.03}$)Ti$_{1.86}$O$_{7.00}$ or Ca$_{0.86}$(Al$_{0.10}$Gd$_{0.05}$Hf$_{0.79}$Pu$_{0.13}$U$_{0.03}$)(Hf$_{0.07}$Ti$_{0.93}$)$_2$O$_{7.00}$. However, no quantitative EPMA results were obtained. After 740 days and dose of 2.6·10^{18} alpha-decays/g the zirconolite had become X-ray amorphous. Calculated bulk density decrease for different samples was from 4.9 to 7.1%. After zirconolite amorphisation all ceramic samples remained physically intact with no evidence of

microcracking. It was reported that "dissolution rate was indepen-
dent of the amount of radiation damage" during a single-pass flow-
through test (SPTF) at 90°C and pH 2.

Synthesis of ^{238}Pu-doped ceramic based on zirconolite of the following composition: $Ca_{0.87}Pu_{0.13}ZrTi_{1.73}Al_{0.23}O_7$, was reported by
French researchers (Advocat *et al.*, 2004). This material contains 10
wt.% PuO_2 (enriched to 93.56 at.% with ^{238}Pu). A radiation damage
study of these samples is in progress.

6.2.9. *Garnet ceramic*

Ceramics based on ^{244}Cm-doped ferrate garnet with calculated average formula $Ca_{1.5}Gd_{0.908}Cm_{0.092}Th_{0.5}ZrFe_4O_{12}$ have been synthesised and studied in RIAR using an IGEM formulation (Lukinykh
et al., 2008). Bulk density was 4.76 g/cm^3. ^{244}Cm content was 2
wt.%, however EPMA of the garnet phase was not carried out. XRD
showed the ceramic sample contained an admixture of ThO_2 and
other unidentified minor phases. The garnet phase had become X-
ray amorphous after a cumulative dose of $1.6 \cdot 10^{18}$ alpha-decays/g or
$76 \cdot 10^{23}$ alpha-decays/m^3. Under self-irradiation by $1.0 \cdot 10^{18}$ alpha-
decays/g the unit-cell parameter of garnet changed from 12.652(2)
to 12.69(1) Å. Geometrical measurements did not show noticeable
change of ceramic density when the garnet phase became amorphous.
Static leach tests (at 90°C, in distilled water, for 3, 7 and 14 days)
revealed amorphisation caused increased Cm leach rate 4–4.5 times
compared to a crystalline sample (from $1.4 \cdot 10^{-3}$ to $5.8 \cdot 10^{-3}$ g/m^2
day on the 14th day).

6.2.10. *Silicate-apatite ceramic*
and chlorine-apatite powder

Samples of ceramic based on ^{244}Cm-doped silicate apatite,
$Ca_2(Nd,Cm)_8(SiO_4)_6O_2$ were made at PNNL in the early 1980s
(Weber, 1982; 1983). Doping level was 2.3 mol.% Cm_2O_3 (62%
^{244}Cm). The apatite had become X-ray amorphous by a cumulative dose of $(110–120) \cdot 10^{23}$ alpha-decays/m^3. No microcracking as
a result of apatite amorphisation was observed under the optical

microscope. Density decrease after cumulative dose $110 \cdot 10^{23}$ alpha-decays/m^3 was close to 8%.

Powdered ^{238}Pu- and ^{241}Am-doped samples of chlorapatite, $Ca_5(PO_4)_3Cl$, and spodiosite, $Ca_2(PO_4)Cl$, were fabricated at PNNL (Metcalfe *et al.*, 2004). The starting precursor contained 3.8 wt.% PuCl$_3$ and 0.2 wt.% AmCl$_3$. SEM and EPMA were not carried out. No damage over the first 532 days at calculated dose of $0.6 \cdot 10^{18}$ alpha-decays/g was detected by XRD.

6.3. Main Points from Self-irradiation/radiation Damage Studies

Numerous results obtained by international teams on the behaviour of actinide-doped crystalline materials under self-irradiation reveal:

— zirconia and zirconate phases with cubic (fluorite of pyrochlore) structure are the most resistant to radiation damage;
— the stability of zirconia-based actinide solid solutions to radiation damage was confirmed but the behaviour of actinide-doped zirconates with pyrochlore structure under self-irradiation (and structure change to fluorite-type) is so far unclear;
— even heavily damaged materials may remain chemically durable, although in general, self-irradiation increases actinide leach rates;
— even heavily damaged materials may remain mechanically stable;
— swelling does not always cause microcracking of ceramic material, although crack formation is typically observed during material amorphisation.
— possible destruction of actinide solid solutions under self-irradiation into new-formed (and possibly, chemically weak) phases requires further detailed research using techniques such as SEM, CL, EPMA and TEM.

References

Advocat T., Jorion F., Marcillat T., Leturcq G., Deschanels X., Boubals J.M., Bojat L., Nivet P. and Peuget S. (2004) Fabrication of $^{239/238}$Pu-zirconolite ceramic pellets by natural sintering, *Scientific Basis for Nuclear Waste Management XXVII, Mater. Res. Soc. Symp. Proc.* **807**, 267–272.

Burakov B.E., Anderson E.B., Zamoryanskaya M.V., Nikolaeva E.V., Strykanova E.E. and Yagovkina M.A. (2001) Investigation of zircon/zirconia ceramics doped with ^{239}Pu and ^{238}Pu. CD-ROM, *Proc. Int. Conf. Global'01*, Paris, France, 9-13/09/2001, paper 006.

Burakov B.E. and Anderson E.B. (2002) Summary of Pu ceramics developed for Pu immobilization (B506216, B512161), in *Review of Excess Weapons Disposition: LLNL Contract Work in Russia*, eds. Jardine L.J. and Borisov G.B. Proc. 3rd Annual Meeting for Coordination and Review of LLNL Work, St. Petersburg, Russia. UCRL-ID-149341, pp. 265–270.

Burakov B.E., Anderson E.B., Yagovkina M., Zamoryanskaya M. and Niko-laeva E. (2002a) Behavior of ^{238}Pu-doped ceramics based on cubic zirconia and pyrochlore under radiation damage, *J. Nucl. Sci. Technol.* Suppl. 3, pp. 733–736.

Burakov B.E., Anderson E.B., Zamoryanskaya M.V., Yagovkina M.A. and Nikolaeva E.V. (2002b) Synthesis and characterization of cubic zirconia, $(Zr,Gd,Pu)O_2$, doped with ^{238}Pu, *Scientific Basis for Nuclear Waste Management XXV, Mater. Res. Soc. Symp. Proc.* **713**, 333–336.

Burakov B.E., Yagovkina M.A., Garbuzov V.M., Kitsay A.A. and Zirlin V.A. (2004a) Self-irradiation of monazite ceramics: contrasting behavior of $PuPO_4$ and $(La,Pu)PO_4$ doped with Pu-238, *Scientific Basis for Nuclear Waste Management XXVIII, Mater. Res. Soc. Symp. Proc.* **824**, 219–224.

Burakov B.E., Yagovkina M.A., Zamoryanskaya M.V., Kitsay A.A., Garbuzov V.M., Anderson E.B. and Pankov A.S. (2004b) Behavior of ^{238}Pu-doped cubic zirconia under self-irradiation, *Scientific Basis for Nuclear Waste Management XXVII, Mater. Res. Soc. Symp. Proc.* **807**, 213–217.

Burakov B.E., Smetannikov A.Ph. and Anderson E.B. (2006) Investigation of natural and artificial Zr-silicate gels, in *Scientific Basis for Nuclear Waste Management XXIX, Mater. Res. Soc. Symp. Proc.* **932**, 1017–1024.

Burakov B.E., Yagovkina M.A., Zamoryanskaya M.V., Garbuzov V.M., Zirlin V.A. and Kitsay A.A. (2008) Self-irradiation of ceramics and single crystals doped with Pu-238: Summary of 5 years of research of the V.G. Khlopin Radium Institute, *Scientific Basis for Nuclear Waste Management, Mater. Res. Soc. Symp. Proc.* **1107**, 381–388.

Burnstall R.S. (1979) FISPIN a computer code for nuclide inventory calculation, Technical report ND-R-328R.

Busker G., Chroneos A., Grimes R.W. and Chen I.W. (1999) Solution mechanisms of dopant oxides in yttria, *J. Am. Ceramic Soc.* **82**, 1553–1559.

Chrosch J., Colombo M., Malcherek T., Salje E.K.H., Groat L.A. and Bismayer U. (1998) Thermal annealing of radiation damaged titanite, *Am. Miner.* **83**, 1083–1091.

Clinard Jr. F.W., Hobbs L.W., Lands C.C., Peterson D.E., Rohr D.L. and Roof R.B. (1982) Alpha decay self-irradiation damage in ^{238}Pu-substituted zircono-lite, *J. Nucl. Mater.* **105**, 248–256.

Degueldre C., Heimgartner P., Ledergerber G., Sasajima N., Hojou K., Muromura T., Wang L., Gong W. and Ewing R. (1997) Behaviour of zirconia based fuel material under Xe irradiation, *Mater. Res. Soc. Symp. Proc.* **439**, 625–632.

Demkowicz M.J., Hoagland R.G. and Hirth J.P. (2008) Interface structure and radiation damage resistance in Cu-Nb multilayer nanocomposites, *Phys. Rev. Lett.* **100**, 136102.

Exharos G.J. (1984) Induced swelling in radiation damaged $ZrSiO_4$, *Nucl. Instrum. Meth. Phys. Res.* **B1**, 538–541.

Ewing R.C., Chakoumakos B.C., Lumpkin G.R. and Murakami T. (1987) The metamict state, *MRS Bulletin*, May 16/June 15, 58–66.

Ewing R.C., Weber W.J. and Lian J. (2004) Nuclear waste disposal — pyrochlore ($A_2B_2O_7$): Nuclear waste form for the immobilization of plutonium and "minor" actinides, *J. Appl. Phys.* **95**, 5949–5971.

Farnan I., Cho H. and Weber W.J. (2007) Quantification of actinide α-radiation damage in minerals and ceramics, *Nature* **445**, 190–193.

Ferry C., Poinssot C., Cappelaere C., Desgranges L., Jegou C., Miserque F., Piron J., Roudil D. and Gras J. (2006) Specific outcomes of the research on the spent fuel long-term evolution in interim dry storage and deep geological disposal, *J. Nucl. Mater.* **352**, 246–253.

Fortner J.A., Badyal Y., Price D.C.L., Hanchar J.M. and Weber W.J. (1999) Structural analysis of a completely amorphous ^{238}Pu-doped zircon by neutron diffraction, *Mater. Res. Soc. Symp. Proc.* **540**, 349–353.

Geisler T., Burakov B., Yagovkina M., Garbuzov V., Zamoryanskaya M., Zirlin V. and Nikolaeva L. (2005) Structural recovery of self-irradiated natural and ^{238}Pu-doped zircon in an acidic solution at $175°C$. *J. Nucl. Mater.* **336**, 22–30.

Grimes R.W. and Catlow C.R.A. (1991) The stability of fission products in uranium dioxide, *Phil. Trans. Roy. Soc. Lond.* **335**, 609–634.

Hambley M.J., Dumbill S., Maddrell E.R. and Scales C.R. (2008) Characterisation of 20-year-old ^{238}Pu-doped Synroc C, *Scientific Basis for Nuclear Waste Management*, *Mater. Res. Soc. Symp. Proc.* **1107**, 373–380.

Hanchar J.M., Burakov B.E., Anderson E.B. and Zamoryanskaya M.V. (2003) Investigation of single crystal zircon, $(Zr,Pu)SiO_4$, doped with ^{238}Pu, *Scientific Basis for Nuclear Waste Management XXVI*, *Mater. Res. Soc. Symp. Proc.* **757**, 215–225.

Holmes-Siedle A.G. and Adams L. (2002) *Handbook of Radiation Effects* (Oxford University Press, England).

Kachalov M.B., Poluektov P.P. and Ozhovan M.I. (1987) Role of inhomogeneities in the fracturing of matrices with radioactive waste, *Soviet Atomic Energy* **63**, 782–784.

Kotomin E.A., Mastrikov Yu.A., Rashkeev S.N. and Van Uffelen P. (2009) Implementing first principles calculations of defect migration in a fuel performance code for UN simulations, *J. Nucl. Mater.* **393**, 292–299.

Lee W.E., Jenkins M.L. and Pells G.P. (1985) The influence of Helium doping on the damage microstructure of heavy-ion irradiation of α-Al_2O_3, *Phil. Mag.* **A51**(5), 639–659.

Lee W.E. (1983) A TEM study of heavy-ion radiation damage in α-Al_2O_3 with and without Helium preimplantation. Ph.D. Thesis, University of Oxford, UK.

Lehmann C. (1977) *Interaction of Radiation with Solids and Elementary Defects Production* (North-Holland, Amsterdam).

Leslie M. (1982) Program CASCADE, description of data sets for use in crystal defect calculations, Technical Report, SERC Daresbury Laboratory Report DL/SCI/TM31T.

Lian J., Wang L.M., Ewing R.C., Yudintsev S.V. and Stefanovsky S.V. (2005) Ion-beam-induced amorphization and order-disorder transition in the murataite structure, *J. Appl. Phys.* **97**, N113536.

Lukinykh A.N., Tomilin S.V., Lizin A.A. and Bychkov A.V. (2002) Investigation of radiation and chemical stability of titanate ceramics intended for actinides disposal (B501111), in *Review of Excess Weapons Disposition: LLNL Contract Work in Russia*, eds. Jardine L.J. and Borisov G.B., Proc. 3rd Annual Meeting for Coordination and Review of LLNL Work, UCRL-ID-149341, St. Petersburg, Russia.

Lukinykh A.N., Tomilin C.V., Lizin A.A. and Livshits T.C. (2008) Radiation and chemical durability of artificial ceramic based on ferrite garnet, *Radiokhimia* **50**(4), 375–379.

Meldrum A., Wang L.M. and Ewing R.C. (1997a) Electron-irradiation-induced phase segregation in crystalline and amorphous apatite: A TEM study, *Am. Miner.* **82**, 858–869.

Meldrum A., Boatner L.A. and Ewing R.C. (1997b) Electron-irradiation-induced nucleation and growth in amorphous $LaPO_4$, $ScPO_4$ and zircon, *J. Mater. Res.* **12**, 1816–1827.

Meldrum A., Boatner L.A. and Ewing R.C. (1997c) Displacive radiation effects in the monazite- and zircon-structure orthophosphates, *Phys. Rev. B* **56**, 13805–13814.

Meldrum A., Boatner L.A., Zinkle S.J., Wang S.X., Wang L.M. and Ewing R.C. (1999) Effects of dose rate and temperature on the crystalline-to metamict transformation in the ABO_4 orthosilicates, *Canad. Miner.* **37**, 207–221.

Metcalfe B.L., Donald I.W., Scheele R.D. and Strachan D.M. (2004) The immobilization of chloride-containing actinide waste in a calcium phosphate ceramic host: Ageing studies, *Scientific Basis for Nuclear Waste Management XXVIII*, *Mater. Res. Soc. Symp. Proc.* **824**, 255–260.

Misra A., Demkowicz M.J., Zhang X. and Hoagland R.G. (2007) The radiation damage tolerance of ultra-high strength nanolayered composites, *J. Metals* **59**, 62–65.

Murakami T., Chakoumakos B.C., Ewing R.C., Lumpkin G.R. and Weber W.J. (1991) Alpha-decay event damage in zircon, *Am. Miner.* **76**, 1510–1532.

Nixon W. and Macinnes D.A. (1981) A model for bubble diffusion in uranium dioxide, *J. Nucl. Mater.* **101**, 192–199.

Ojovan M.I. and Poluektov P.P. (2001) Surface self-diffusion instability in electric fields, *Mater. Res. Symp. Proc.* **648**, P3.1.1–P3.1.6.

Parfitt D.C. and Grimes R.W. (2009) Predicting the probability for fission gas resolution into uranium dioxide, *J. Nucl. Mater.* **392**, 28–34.

Parfitt D.C. and Grimes R.W. (2008) Predicted mechanisms for radiation enhanced helium resolution in uranium dioxide, *J. Nucl. Mater.* **381**, 216–222.

Pichot E., Dacheux N., Emery J., Chaumont J., Brandel V. and Genet M. (2001) Preliminary study of irradiation effects on thorium phosphate-diphosphate, *J. Nucl. Mater.* **289**, 219–226.

Polezhaev Yu.M. (1974) On the mechanism of metamictization of minerals under the action of autoradiation, *Geokhimiya* **11**, 1648–1652.

Raison P.E., Haire R.G. and Assefa Z. (2002) Fundamental aspects of Am and Cm in zirconia-based materials: Investigations using X-ray diffraction and Raman spectroscopy, *J. Nucl. Sci. Tech.* Suppl. 3, 725–728.

Sickafus K.E., Matzke Hj., Hartmann Th., Yasuda K., Valdez J.A., Chodak III P., Nastasi M. and Verrall R.A. (1999) Radiation damage effects in zirconia, *J. Nucl. Mater.* **274**, 66–77.

Sickafus K.E., Minervini L., Grimes R.W., Valdez J.A., Ishimaru M., Li F., McClellan K.J. and Hartmann T. (2000) Radiation tolerance of complex oxides, *Science* **289**, 748–751.

Sickafus K.E., Grimes R.W., Valdez J.A., Cleave A., Tang M., Ishimaru M., Corish S.M., Stanek C.R. and Uberuaga B.P. (2007) Radiation-induced amorphization resistance and radiation tolerance in structurally related oxides, *Nature Materials* **6**, 217–223.

Soulet S., Carpena J., Chaumont J., Kaitasov O., Ruault M.O. and Krupa J.C. (2001) Simulation of the α-annealing effect in apatitic structures by He-ion irradiation: Influence of the silicate/phosphate ration and of the OH$^-$/F$^-$ substitution. *Nucl. Instrum. Meth. Phys. Res.* **B184**, pp. 383–390.

Strachan D.M., Scheele R.D., Buchmiller W.C., Vienna J.D., Sell R.L. and Elovich R.J. (2000) Preparation of ^{238}Pu-ceramics for radiation damage experiments, Report PNNL-13251 under contract DE-A-C06-76RLO 1830.

Strachan D.M., Scheele R.D., Kozelisky A.E., Sell R.L., Todd Schael H., O'Hara M.J., Brown C.F. and Buchmiller W.C. (2002) Radiation damage in titanate ceramics for plutonium immobilization, *Scientific Basis for Nuclear Waste Management XXV, Mater. Res. Soc. Symp. Proc.* **713**, 461–468.

Strachan D.M., Scheele R.D., Icenhower J.P., Buck E.C., Kozelisky A.E., Sell R.L., Elovich R.J. and Buchmiller W.C. (2004) Radiation damage effects in candidate ceramics for plutonium immobilization: Final Report, PNNL-14588, Richland, Washington 99352.

Strachan D.M., Scheele R.D., Buck E.C., Icenhower J.P., Kozelisky A.E., Sell R.L., Elovich R.J. and Buchmiller W.C. (2005) Radiation damage effects in candidate titanates for Pu disposition: Pyrochlore, *J. Nucl. Mater.* **345**, 109–135.

Strachan D.M., Scheele R.D., Buck E.C., Kozelisky A.E., Sell R.L., Elovich R.J. and Buchmiller W.C. (2008) Radiation damage effects in candidate titanates for Pu disposition: Zirconolite, *J. Nucl. Mater.* **372**, 16–31.

Thompson M.W. (1969) *Defects and Radiation Damage in Metals*, (Cambridge University Press, UK).

Trachenko K., Dove M.T. and Salje E.K.H. (2002) Structural changes in zircon under α-decay irradiation, *Phys. Rev. B* **65**, 180102.

Utsunomiya S., Wang L.M., Yudintsev S. and Ewing R.C. (2002) Ion irradiation-induced amorphization and nano-crystal formation in garnets, *J. Nucl. Mater.* **303**, 177–187.

Volkov Yu.F., Lukinykh A.N., Tomilin S.V. and Bychkov A.V. (2001) Investigation of U.S. titanate ceramics radiation damage due to ^{238}Pu alpha-decay [B501111], in *Immobilization of Excess Weapons Plutonium in Russia: A Review of LLNL Contract Work*, eds. Jardine L.J. and Borisov G.B., Proc. Meeting for Coordination and Review of Work, UCRL-1D-143246 St. Petersburg, Russia, (2000), pp. 196–207.

Wang S.X., Begg B.D., Wang L.M., Ewing R.C., Weber W.J. and Govidan Kutty K.V. (1999) Radiation stability of gadolinium zirconate: A waste form for plutonium disposition, *J. Nucl. Mater.* **14**, 4470–4473.

Wang L.M. and Weber W.J. (1999) Transmission electron microscopy study of ion-beam-induced amorphization of $Ca_2La_8(SiO_4)_6O_2$, *Phil. Mag. A* **79**, 237–253.

Wang L.M., Wang S.X. and Ewing R.C. (2000) Amorphization of cubic zirconia by caesium-ion implantation, *Phil. Mag. Lett.* **80**, 341–347.

Was G.S. (2007) *Fundamentals of Radiation Materials Science: Metals and Alloys* (Springer-Verlag, Germany).

Weber W.J., Turcotte R.P., Bunnell L.R., Roberts F.P. and Westsik J.H. (1979) Radiation effects in vitreous and devitrified simulated waste glass, in *Ceramics in Nuclear Waste Management*, eds. Chikalla T.D. and Mendel J.E., CONF-790420, Nationals Technical Information Service, Springfield, Virginia, pp. 294–299.

Weber W.J. (1982) Radiation damage in rare-earth silicate with the apatite structure, *J. Am. Ceram. Soc.* **65**(11), 544–548.

Weber W.J. (1983) Radiation-induced swelling and amorphization $Ca_2Nd_8(SiO_4)_6O_2$, *Radiat. Effect* **77**, 295–308.

Weber W.J., Wald J.W. and Matzke Hj. (1986) Effect of self-radiation damage in Cm-doped $Gd_2Ti_2O_7$ and $CaZrTi_2O_7$, *J. Nucl. Mater.* **138**, 196–209.

Weber W.J. (1991) Self-radiation damage and recovery in Pu-doped zircon. *Radiat. Eff. Defect Solid* **115**, 341–349.

Weber W.J., Ewing R.C. and Wang L.-M. (1994) The radiation-induced crystalline-to-amorphous transition in zircon, *J. Mater. Res.* **9**, 688–698.

Weber W.J., Ewing R.C., Catlow C.R.A., Diaz de la Rubia T., Hobbs L.W., Kinoshita C., Matzke Hj., Motta A.T., Nastasi M., Salje E.K.H., Vance E.R. and Zinkle S.J. (1998) Radiation effects in crystalline ceramics for the immobilization of high-level nuclear waste and plutonium, *J. Mater. Res.* **13**, 1434–1484.

Yudintsev S.V., Lukinykh A.N., Tomilin S.V., Lizin A.A. and Stefanovsky S.V. (2009) Alpha-decay induced amorphization in Cm-doped Gd_2TiZrO_7, *J. Nucl. Mater.* **385**, 200–203.

Zamoryanskaya M.V., Burakov B.E., Bogdanov R.R. and Sergeev A.S. (2002) A cathodoluminescence investigation of pyrochlore, $(Ca,Gd,Hf,U,Pu)_2Ti_2O_7$, doped with ^{238}Pu and ^{239}Pu, *Scientific Basis for Nuclear Waste Management XXV, Mater. Res. Soc. Symp. Proc.* **713**, 481–485.

Zamoryanskaya M.V. and Burakov B.E. (2004) Electron microprobe investigation of Ti-pyrochlore doped with Pu-238, *Scientific Basis for Nuclear Waste Management XXVIII, Mater. Res. Soc. Symp. Proc.* **824**, 231–236.

Chapter 7

What is the Future?

If the present trend of the increased use of actinides is to continue then durable immobilisation host materials are required. Crystalline matrices are currently the most promising.

7.1. Safety Issues

Actinides are characterised by several unique properties which are attractive for industrial application in many fields, including nuclear weapons and fuel, sources of alpha-irradiation, superconductors and "nuclear" batteries. Further study of actinides promises new actinide applications, but at the same time most actinide isotopes are highly dangerous and their application must be strictly limited and controlled (Table 1.1.2). Safety of actinide use therefore is a careful balance between civilian application and serious threats such as nuclear proliferation and terrorism and adverse impact on human health and the natural environment. It is difficult to predict the future of the nuclear industry but in any case there is no doubt that actinides already play and have the potential to take on an even greater role in the future. We must develop a proper approach to actinide use and disposal while avoiding any risk of actinide involvement in illegal

190   *Crystalline Materials for Actinide Immobilisation*

activity. There are some important details to be considered in this approach:

— international agreements on peaceful use of nuclear energy, nuclear disarmament and nonproliferation;
— development of chemically- and mechanically-durable forms of actinides for any use and disposal;
— use of minimum amounts of actinide in any form for broad use, e.g., sealed sources, "nuclear" batteries;
— development of reliable systems of actinide waste management that will potentially include the use of transmutation and deep geological disposal.

7.2. Burning (Transmutation Option)

Although complete burning (transmutation) of dangerous actinide isotopes in nuclear reactors is a popular subject of broad scientific discussion, no cost or energy-effective transmutation technology has been discovered. Current possibilities of partial transmutation are limited by the following options:

— improvement of reactor design in order to increase fuel burn-up (for uranium oxide and MOX fuel);
— development of ceramic nuclear fuel (see Chap. 2);
— development of a new generation of nuclear reactors to use uranium oxide or MOX fuel with essential admixtures of minor actinides. An alternative option is the use of a thorium fuel cycle that avoids significant accumulation of minor actinides.

7.3. Disposal of Waste Actinides

All current approaches to the disposal of highly radioactive actinide wastes are based on a multi-barrier concept. The geological barrier is considered the most important and reliable. Before emplacement in caverns in deep geological host-rock the solid wasteforms should be packed into canisters and (or) containers surrounded by buffer materials (e.g., bentonite). Various rocks (salt, granite, clay, argillites) can be used depending on geological conditions as well as the local social situation and public acceptance in any particular country. Repository designs include the use of shafts and tunnels at depths of about 500–1000 metres as well as potentially very deep (more than 3,000 metres) boreholes. Different approaches to actinide disposal may be characterised into two groups:

(i) Engineering approaches *and*
(ii) Geochemical approaches.

The engineering approach considers waste repository design as the main basis of long-term safety. Justification of repository design includes:

— selection of host-rock with low permeability for underground water;
— identification of optimal repository depth;
— the development of reliable waste container and near-field materials such as buffer and backfill.

Such an approach allows disposal of spent fuel, which is not a particularly stable wasteform.

The geochemical approach (accepting all engineering developments) considers disposal of long-lived actinides as a process full of uncertainties. According to this approach any geological environment disturbed by construction of a repository supports degradation of engineering barriers much faster than might be predicted. Radiation fields in the near field and self-irradiation of waste matrices may speed up the processes of chemical corrosion and radionuclide migration via water-soluble species and colloids. Development of durable wasteforms for actinide disposal is therefore extremely important. Spent fuel cannot be considered an appropriate wasteform. Justification of wasteforms is based not only on chemical durability and resistance to radiation damage but also geochemical compatibility with the host-rock (Fig. 1.4.2).

192   *Crystalline Materials for Actinide Immobilisation*

7.4. Performance of Actinides in Disposal Environment

The behaviour of actinides under the conditions of the disposal environment is a subject of long-term detailed research. Current knowledge is based mainly on the study of some analogues: natural deposits of uranium and thorium, nuclear test sites, sites of nuclear accidents (for example, Chernobyl in Ukraine; Russia's Mayak-site), sorption-migration experiments using tracer amount of actinides (Grimsel-site in Switzerland) and laboratory-scale tests on leaching of actinides from glass and ceramic wasteforms. However, all these important data cannot compensate for the lack of information concerning migration and retardation of long-lived, highly radioactive actinide isotopes in geological formations over many thousands of years. A possible way to solve this problem is building pilot-scale repositories provided with permanent monitoring systems for actinide migration. It would be useful also to test full-size waste packages loaded with particular amounts of actinides under the conditions of a real geological environment. Such work requires licensing and public acceptance.

7.5. Conclusions

Throughout this book we have highlighted not only the value but also the dangers, associated with actinide-containing materials. We have focused on actinide immobilisation in potential wasteforms but new fuel cycles such as those for fast neutron reactors and the possible advent of accelerator-driven systems, require a different approach to actinide immobilisation. New management methods for high and intermediate-level nuclear wastes are under consideration, notably partitioning-transmutation (P&T) and partitioning-conditioning (P&C), where the prime objective is to separate long-lived radionuclides from short-lived ones. Potential P&T targets were described in Sec. 2.5. P&T aims to change hazardous long-lived radionuclides into significantly shorter-lived radionuclides. P&C aims to process and dispose of long-lived radionuclides separately from shorter-lived radionuclides which could themselves be immobilised in wasteforms (probably vitreous) destined for near surface disposal. Both P&T and P&C rely on materials being developed which are capable of hosting large amounts of actinides and which are also durable. We believe the crystalline materials and processes described here have the potential for significant future application in these and many other uses.

Index

Accessory minerals, 36–39, 46

Actinium, 2, 5, 9

Admixture, 11, 14, 21–23, 27, 30, 34, 35, 40, 41, 43, 44, 46, 47, 66, 76, 83, 84, 90, 91, 112, 117, 127, 141, 168, 180, 190

AGR, 57

Allotropes, 8

Alpha radiation, 14, 15, 17, 68, 135, 136, 139, 162, 164

Alteration, 25, 38, 43, 84, 145, 147, 148, 151

Alteration tests, 145, 147

Americium, 3, 7, 9, 17, 31, 33–36, 63, 64, 81

Amorphisation, 82, 87, 158, 161–164, 167, 176, 178–181

Apatite, 37, 44, 45, 74, 77, 88, 89, 121

Baddeleyite, 42, 43

Be, 167, 168, 170, 174, 177

Be-window, 167, 168, 170, 174, 177

Becquerel, 1, 5

Berkelium, 4, 36

Boiling Water Reactor (BWR), 57

Brannerite, 24, 25

Burning, 28, 57–59, 190

Calcite, 24

Californium, 4, 9, 36, 79, 177

Cathodoluminescence, 66, 68, 80, 136, 139–141, 159, 177

Ceramic nuclear fuel, 58, 190

Ceramic wasteform, 37, 39, 73–75, 80, 84, 88, 95, 100, 128, 145, 151, 192

Chadwick, 5

Chernobyl lava, 75, 76, 138

Chlorapatite, 80, 81, 89, 116, 181

Co-precipitation, 115, 116, 128

Coffinite, 23, 24

Cold Crucible Melting (CCM), 79, 94, 97, 98, 121, 125, 128

Crystallisation, 51, 74, 95, 114, 120, 122, 125, 128, 162

Cubic zirconia, 43, 59, 60, 69, 75, 77–79, 81, 82, 92–94, 99, 100, 143, 150, 161, 170, 171

Curie, 5–7

Curium, 4, 7, 9, 35, 36

Depleted Uranium (DU), 15

Disposal, 1, 20, 21, 33, 35, 37–39, 60, 61, 67, 69, 77, 82, 145, 151, 189–193

Einsteinium, 4

EMPA, 134–140, 142, 166, 167, 169, 170, 175, 179–181

Fermi, 5

Fermium, 4

Fission products, 3, 26, 29, 59

Fluorapatite, 89, 115, 121, 161, 165

Flux method, 24, 32, 34, 73, 124–126, 128, 168

196   *Crystalline Materials for Actinide Immobilisation*

Gamma radiation, 15, 64
Garnet, 37, 47–49, 78, 83, 94–97, 100,
    116, 121, 122, 124, 125, 141, 161,
    165, 180
Gel, 23, 50, 51, 90, 111–115, 162

Hafnia, 89, 90, 92, 94, 99, 100, 166,
    168
Hafnon, 38, 40, 79, 89, 90, 99, 100,
    164, 166–168
Hibonite, 69
High Level Waste (HLW), 29, 74–77,
    80, 83, 84, 88, 112
Hot Isostatic Pressing (HIP), 118, 128
"Hot" particles, 21, 138
Hot Uniaxial Pressing (HUP), 117,
    118, 128

IAEA, 19, 21, 73, 146, 147
Immobilisation, 1, 19–21, 33, 34, 36,
    37, 65, 73–80, 84, 86, 88–90, 92, 94,
    95, 99, 118, 119, 128, 145, 165, 189,
    192
Inert Matrix Fuels (IMFs), 59, 68,
    100, 159
Ion irradiation, 77, 78, 163–165

Kosnarite, 49, 50, 80, 88, 99

Lanthanides, 7, 8, 35, 36
Leach tests, 78, 80, 146–151, 167, 175,
    176, 180
Light Water Reactors (LWR), 26, 28,
    29, 77
Luminescence, 65–67, 94, 124

Magnox, 26
Mechanical durability, 21, 25, 40, 49,
    58, 66, 82, 87, 88, 90, 92, 99, 100,
    143, 164, 167, 190
Melting, 14, 43, 78, 79, 84, 86, 89,
    92–98, 120–122, 128
Melting-crystallisation, 120, 122, 128
Mendelevium, 4
Metamict, 23, 24, 40–43, 46, 47, 51,
    80, 82, 159, 161

Minor Actinides (MA), 1, 28, 29, 33,
    35, 58, 60, 159, 190
Mixed Oxide (MOX), 16, 30, 57–61,
    69, 117, 118, 128, 159, 190
Monazite, 3, 37, 38, 41, 74, 76, 77, 81,
    82, 87–89, 99, 115, 116, 124, 125,
    128, 141, 142, 161, 163, 164, 171,
    172
Murataite, 49, 78–80, 99, 121, 161,
    164, 165

Neptunium, 3, 6, 9, 33, 34, 94
Normalised mass loss, 80, 149, 150,
    171, 175, 178
NZP, 80, 88, 99

Oklo (natural reactor), 27
Optical couplers, 66
Optical microscopy, 136, 140, 142,
    143, 152, 173

Perovskite, 46, 47, 76, 78, 79, 83,
    94–97, 99, 100, 121, 122, 179
Phosphorescence, 1
Pitchblende, 2, 3, 5, 21, 23, 24
Placer deposit, 37, 49, 89
Plutonium, 3, 6–9, 18, 21, 26, 28–35,
    57–61, 63, 65, 76–84, 87–89, 92, 94,
    140
Polonium, 5
Pressing, 74, 84–86, 94, 117–120, 128
Pressing-sintering, 118, 128
Pressurised Water Reactor (PWR),
    26, 57, 59, 77
Protactinium, 2, 5, 9
Pyrite, 24
Pyrochlore, 37, 45–47, 68, 74, 76–86,
    118, 141, 150, 158, 161, 163–165,
    173, 175–178, 181
Pyrophoric, 15

Radiation damage, 25, 30, 43, 44, 49,
    67, 78–82, 135, 137, 140, 143,
    157–166, 180, 181, 191
Radiation effects, 81–83, 157, 162

Radiolysis, 19
Radiotoxicity, 14, 17, 19, 21, 30, 35, 61, 69
Radium, 5, 7, 14, 16, 32, 34, 63–66, 68, 75, 85, 86, 91, 93, 96
Radon, 5, 77, 79, 178
Rare Earth Elements (REE), 37, 77, 88, 99
RBMK reactors, 26, 29
Repository, 38, 39, 59, 84, 151, 191
Roentgen, 5
Rutherford, 5

Safety issues, 189
Sealed source, 63–65, 133, 190
Self sustaining high, 80, 84, 94, 97, 122, 128
Self-glowing materials, 20, 65, 67, 83
Self-irradiation, 32, 37, 38, 40, 43, 51, 66, 80–82, 86, 91, 94, 139, 141–144, 151, 157, 159, 162–164, 166, 167, 169, 171–173, 175–181, 191
SEM, 76, 81, 85, 86, 93, 96, 97, 124, 134–137, 139, 140, 166, 170, 175, 176, 181
Single crystal growth, 123
Single crystals, 13, 32, 34, 39, 73–75, 80–82, 89, 90, 121, 123, 125, 127–129, 136, 137, 139, 140, 142–144, 160, 161, 164, 166, 168, 170, 172, 174, 176
Smoke detector, 17, 35, 61–63
Sol-gel, 90, 91, 111, 114, 128
Spent Nuclear Fuel (SNF), 23, 28–30, 33–36, 59
Synroc, 73–75, 77, 78, 80, 82–84, 99, 100, 117, 122, 179

Tazheranite, 43, 92
TEM, 93, 114, 115, 159, 165, 173, 178, 181
Temperature reaction, 84, 94, 97, 122
Thorianite, 21–23, 37
Thorite, 2, 23, 24, 37
Thorium, 1, 2, 5, 8, 9, 11, 14, 15, 21, 25, 26, 41, 60, 77, 87, 115, 150, 190, 192
Thorium oxide (thoria), 11
Titanates, 25, 46, 74, 76, 78, 79, 81–84, 150, 163, 175, 177
TPD, 77, 87, 88, 99, 115, 150, 165
Transmutation, 15, 20, 25, 26, 28, 30, 33, 34, 58, 60, 68, 69, 94, 99, 100, 190, 192
Transmutation targets, 68, 69, 94

Uraninite, 21–23, 37
Uranium, 1–3, 5–9, 14–16, 19, 21–29, 31, 41, 46, 50, 51, 57, 73, 86, 95, 123, 159, 177, 190, 192
Uranium oxide, 2, 3, 5, 22, 23, 27, 28, 57, 123, 190

Xenotime, 37, 43, 44, 66–68, 99, 125, 141, 161
XRD, 24, 32, 133, 134, 143, 151, 159, 166–170, 172–175, 177, 178, 180, 181

Zircon, 3, 37–42, 50, 51, 66–68, 73–77, 80, 82, 83, 89–93, 112, 114, 123–125, 127, 128, 137–144, 150, 160–164, 166–170
Zirconolite, 41, 42, 74, 76–79, 81, 83, 84, 99, 150, 161, 163, 175, 178–180

www.ingramcontent.com/pod-product-compliance
Lightning Source LLC
Chambersburg PA
CBHW050601190326
41458CB00007B/2135